T0206765

The Awa Book of New Zealand Science

Also by Rebecca Priestley

Atoms, Dinosaurs and DNA: 68 Great New Zealand Scientists
(with Veronika Meduna)

The Elegant Universe of Albert Einstein (contrib.)

The Awa Book of
New Zealand Science

Edited by Rebecca Priestley

AWA SCIENCE

First published in 2008 by Awa Press,
16 Walter Street, Wellington, New Zealand

This anthology © Rebecca Priestley 2008
For further copyright information see page 355

The right of Rebecca Priestley to be identified as the editor of
this work in terms of Section 96 of the Copyright Act 1994 is
hereby asserted.

National Library of New Zealand Cataloguing-in-Publication Data
The Awa book of New Zealand science / edited by Rebecca Priestley.
Includes bibliographical references and index.
ISBN 978-0-9582629-9-6
1. Science—New Zealand. I. Priestley, Rebecca.
509.93—dc 22

Book design by Fortyfive Design, Wellington
Typeset by Jill Livestre, Archetype, Wellington
Printed by Everbest Printing Company, China
This book is typeset in Minion

www.awapress.com

For my mother Ruth Brassington
and my father Nigel Priestley

'**Discovery.** There's nothing so exciting or satisfying as making a discovery.'

Mont Liggins

Contents

Introduction

South Wairarapa, February 12, 2005. The opening of Stonehenge Aotearoa. A gale-force wind is whipping across the Wairarapa plains, but instead of retiring inside the flapping walls of the catering marquee the assembled dignitaries, journalists, astronomers and enthusiasts are standing close to the circle of pillars and lintels, gazing anxiously at the sky. In the distance a small black speck hovers to the west of the Rimutaka Range. As the helicopter gets closer we can see it is being tossed vigorously from side to side by the wind gusts. Inside is New Zealand's third, and only living, Nobel laureate, septuagenarian Professor Alan MacDiarmid and his wife Gayl. We're all holding our breath, fearful that we're about to bear witness to the demise of one of our science greats.

Now we can hear the rotors over the sound of the wind; the helicopter is going to make it. Approaching the ring of stones, it circles the area to get the feel of the local wind before landing face-on to the gale, to a collective sigh of relief. The guests of honour emerge and are escorted away to recover before the ceremony starts.

After the official launch I wait for the journalists and fans to disperse, then work up the courage to approach MacDiarmid. I'm more confident dealing with the books and letters of long-deceased scientists than approaching a live one. But I introduce myself, telling him about the anthology of New Zealand science I'm embarking on, and asking if he has written something about his work that I may be able to include. He graciously says he'd be delighted to be included, and refers me to his autobiography on the Nobel Foundation website—an extract of which is included in this book. Later he gives a moving speech, echoing the words of Isaac Newton with his statement that at Stonehenge Aotearoa 'we stand on the shoulders of giants'.

Introduction

New Zealand scientists today also stand on such shoulders, although the stories of many of the giants are only just emerging. A sense of the past is important in any discipline, and for New Zealand science written reflection on the past has, until quite recently, been scant. For decades the only New Zealand scientist about whom you could read a biography was Ernest Rutherford —and there were dozens to choose from. But in the eight years of this century alone we have seen biographies or memoirs of scientists such as naturalist Charles Fleming, geologist Harold Wellman, astronomer Beatrice Hill Tinsley, father and son naturalists G.M. and Allan Thomson, geochemist Brian Mason and agricultural scientist Theodore Rigg. Other science biographies, including one of the evolutionary biologist Allan Wilson, are on their way. I hope this book will also make a positive contribution to the emerging interest in New Zealand's science history. Alan MacDiarmid—himself a giant of New Zealand science—is just one of the scientists (and scientist's families) who reacted enthusiastically, and gave permission for their stories to be included.

In my research over the past four years, I have been struck by the dedication, passion and even obsession of our scientists for their work. Naturalists such as George Hudson and Charles Fleming trace their life's work back to a childhood fascination with the natural world, and included in this book are a piece about collecting insects written by George Hudson when he was nineteen years old, and Charles Fleming's reminiscence of shell-collecting on Auckland beaches as a boy. Science is, at its best, a creative pursuit driven by a search for meaning and a better understanding of the world around us. The creativity of scientists can be as profound as the radical thinking about the origins of humanity, and the universe itself, that earned Allan Wilson and Beatrice Hill Tinsley their international reputations, or as practical as Maurice

Wilkins' ingenuity in adapting his scientific apparatus with a condom and some Plasticine, or Mont Liggins' pioneering experiments on the foetuses of sheep.

In this selection of writings about New Zealand science, I have included pieces by some of the giants, as well as by some of the relatively unsung heroes. While I have aimed to cover different time periods, people and scientific disciplines, my main requirement has been that each piece should be enjoyable and accessible to that mythical creature the 'average reader'—a person with a lively interest in how the world works, but not necessarily a background in science. At the same time, I have tried to include enough meaty scientific detail to satisfy those readers who do have a science background. Although it wasn't my initial plan, it eventually became clear that stories written in the first person, in the scientist's own voice, were the most vibrant and captivating; almost all of the pieces included here are written by the scientists themselves.

The pieces are arranged broadly chronologically. Poems have been selected to give an extra dimension to some of the stories; they appear next to the science or scientist they describe. This choice of sequence is to give a picture of the progression and evolution of science in New Zealand. The first piece in the book, by Maori Studies' head Peter Adds, details the changing theories about the origins and ocean voyages of the Maori. Next, the European naturalists and collectors who visited or lived in New Zealand in the eighteenth and nineteenth centuries describe New Zealand's geology, flora and fauna. Walter Buller tells of his huia-hunting expeditions, Leonard Cockayne tries to say something nice about the 'much-maligned mangroves', and Ferdinand von Hochstetter describes the beauty of the pink and white terraces of Lake Rotomahana. Come the twentieth century, the selection reflects the expansion of science into new fields such as nuclear physics, genetics and astrophysics, and the work of exported New Zealand scientists, many of whom, including our three Nobel Prize-winners, Ernest Rutherford, Maurice Wilkins and Alan MacDiarmid, have made or are making an international impact.

The footnotes in this book are my own; I have left out the original writers' footnotes, some of which were voluminous. Some pieces have been edited to make them shorter and more focused on a single theme—usually that of scientific discovery. In the interests of accessibility, silent editorial changes have been made, such as standardising spelling and correcting grammatical errors. Where passages have been deleted, this is clearly shown with either ellipses (...) or a linking editorial passage. Anyone who would like to read the complete version of an edited work should refer to the original texts, for which full reference details are given at the end of the book.

As indicated by the quote from Mont Liggins on the opening page, the overarching theme of this book is discovery. Researching and editing it has been my own journey of discovery. My initial interest was in twentieth-century physics, particularly New Zealand's nuclear and radiation history, in which the stories of Ernest Rutherford, Ernest Marsden and Athol Rafter loom large. However, as I researched for this anthology I was constantly surprised and thrilled by the rich breadth of work achieved by our scientists in all fields. I came to admire the genius, humanity, and what can often appear to the outsider as sheer barminess, of these men and women who, in their thirst for knowledge, have made a difference to all our lives.

I hope you enjoy reading these pieces as much as I have.

Rebecca Priestley
May 2008

Polynesian Navigation

Long before the arrival of the first European explorers and naturalists, Aotearoa's first inhabitants reached these islands, using a finely honed understanding of the natural world to navigate by the sun and stars, the direction of ocean swells and currents, and evidence of clouds and birds. This fact was not always accepted, however, and it was only in the late twentieth century that scientists became aware of the complexity and extent of the traditional knowledge, or Matauranga Maori, that New Zealand's first peoples carried with them.

In this piece, Peter Adds, head of Tumuaki, the School of Maori Studies at Victoria University, follows the changing theories of the origins and ocean voyages of the Maori.

When Captain Cook was in Tahiti making his observations about the transit of Venus, he was among people who were also profoundly good navigators and sailors, a fact he was slow to recognise. When he and the other European explorers arrived in the Pacific they were bemused, frustrated and even agitated to discover that other peoples had beaten them to these islands. They began wondering how people who appeared to be nothing more than savages had been able to cross the world's biggest ocean, find the islands and settle them. These people didn't have compasses, maps or sextants. Nor did they appear to have very good boats. How on earth had they been able to do it? …

Many of the early theories about the ability of Polynesians to sail and navigate were based on the assumption that, as a race, Polynesians must have degenerated from some former better, more brilliant civilisation. They had to have been good at these arts at one time, the theory went, because they had got to the islands, but after that, in the isolation, they had regressed into the people witnessed by the early European explorers. ...

By the middle of the nineteenth century, however, scholars were starting to take the marine capabilities of Polynesians more seriously. The Polynesians' aural traditions talked about great voyages of discovery, great navigators and great explorers, and some ethnographers started to believe there might be something in these stories. Maybe Polynesians did have a marine technology, a body of knowledge that allowed them to sail to distant places?

By the middle of the twentieth century, an orthodoxy had developed about these ideas. It was, however, an orthodoxy based mostly on assumptions—produced by people such as Te Rangi Hiroa (Sir Peter Buck), Percy Smith, and other ethnographers. By the 1940s this orthodoxy had become very persuasive. It went something like this: Polynesians had originally come from the west, from somewhere in Asia, and had sailed into the Pacific against the winds and the currents. This idea was popularised most importantly by Sir Peter Buck in 1938, when he published a book in the United States called *Vikings of the Sunrise*. In it he presented a romanticised idea of Polynesians' sailing capabilities, as well as his theories about their origin as a race—some of which were, probably rather accidentally, not too far from what we know today to be possibly the case.

The orthodoxy also held that Polynesians had large seaworthy canoes, that they could navigate over very large distances whenever and wherever they felt like it, and that their voyages of settlement and exploration were intentional, and the primary means by which Polynesia had been settled. ...

The orthodoxy was well entrenched in New Zealand folklore and mythology until the mid 1940s, when some people came along who weren't entirely taken with the theory and started to challenge it. The first was the Norwegian adventurer Thor Heyerdahl. In the Pacific, the winds and currents for the most part go from *east to west*, and we had begun to learn from archaeology that the direction of settlement in Polynesia had been from *west to east*. Heyerdahl, who had an archaeological background of sorts, didn't think Polynesians were clever enough, or had the necessary technical expertise, to have sailed against the wind and currents.

Heyerdahl also knew that the kumara, the sweet potato that was the staple diet of many Polynesian peoples, was a South American cultigen, not a South-east Asian one. He put two and two together, and came up with the idea that the Polynesians had come originally from South America.

In 1947, Heyerdahl set out to test that idea by constructing a boat based on ancient South American Indian designs. That boat, the *Kon-Tiki,* was essentially a raft with a large structure on top, one mast, and a couple of sails attached to the mast. It was built largely of local balsa-wood, bundled and tied together, and designed, among other things, to allow the water to flow through. However, when Heyerdahl and his crew attempted to sail out of Callao Harbour in Peru there was a problem: a little thing called the Humboldt Current, which runs swiftly off the western coast of South America in a northerly direction. Caught up in the current, the *Kon-Tiki* was unable to get out of it. It was being swept anywhere but Polynesia. In the end, Heyerdahl had to hire a tugboat to tow the craft 50 miles out to sea across the Humboldt Current.

You might have thought this would negate the experiment, but Heyerdahl continued anyway. The *Kon-Tiki,* being constructed of balsa-wood, soon became waterlogged. And as it sailed in a westerly direction into Polynesia, it became heavier and heavier, and more susceptible to currents, and therefore more difficult to steer. Heyerdahl and his crew did make it to Polynesia eventually, sailing with the wind and going with the currents, but unfortunately they had to pass a whole lot of islands they sighted along the way: the *Kon-Tiki* had become so waterlogged it couldn't be steered.

Finally, about 101 days after they had set out from Peru, the group crash-landed on Raroia, a small island in the Tuamotus. Heyerdahl had proved that if you get towed by a tugboat across the Humboldt Current and are lucky enough, you can crash land on a small tropical island in Polynesia. The scholars of Polynesian history and people who were interested in sailing and navigation did not accept Heyerdahl's theory, but the experiment had wide public appeal. Heyerdahl's book on the voyage sold millions of copies and was translated into around 50 languages. There were television programmes and an Academy Award-winning documentary.

A few years later another doubter came along, this time a New Zealand historian called Andrew Sharp. Sharp was an armchair theorist who, like Heyerdahl, did not believe the early Polynesians would have had the technical ability to sail against the winds and currents. Nor did he think their canoes were good enough, or their knowledge comprehensive enough to navigate between islands over large distances. In his view, therefore, the settlement of Polynesia had been one big accident, a theory he expounded in his 1956 book *Ancient Voyagers in Polynesia*. He argued that people fishing at sea had probably got lost in storms and ended up accidentally arriving on islands, where they subsequently settled. And incidentally, virtually every piece of Pacific real estate that can sustain human life was settled at some time by Polynesians.

But there was also a problem with Sharp's theory, which was that in Polynesia fishing tends to be done solely by men. And if you are going to subscribe to the idea that the settlement of Polynesia was an accident, you have to have some women in there. This was largely overlooked by Sharp.

What Heyerdahl and Sharp did, though, was get people think-ing about what had actually happened. What was needed was some hard data. Enter another New Zealander, a man by the name of David Lewis. Lewis was an explorer, sailor and doctor who had spent some time sailing around the Pacific in his own boat. In Micronesia he had met up with some local sailors and learned—especially from Mau Piailug, a well-known man from the island of Satawal in the Caroline Islands—some of the lost arts of navi-gation. These techniques included navigating by star positions; reading land signs—signs that indicate the proximity of land when at sea in the open ocean; understanding the directional qualities of currents, swells, and the phenomenon called luminescence that appears in the sea at night; and being guided by wind direc-tions and cloud formations. It was a very comprehensive body of knowledge.

In the early 1970s, using these techniques, David Lewis successfully sailed his trimaran from Tahiti to New Zealand, thereby proving that Polynesians had the technical expertise to undertake the amazing voyages of colonisation we know must have happened, and that they could have not only undertaken voyages to new homelands, but turned around and sailed home again relatively easily had they wanted to.

Just how miraculous those voyages were was shown in an ingenious computer simulation carried out in England in 1972 by three meteorologists, Michael Levison, Gerard Ward and John Webb. These men gathered a vast amount of Pacific weather information, including currents, winds and storm conditions, put it together in their supercomputer with data on canoe speeds at different wind speeds and ran a series of over 100,000 simulated voyages. Their aim was to find out what would happen to canoes leaving various islands if they were left to drift.

They found it would be virtually impossible to drift accidentally to many of the islands settled by Polynesians. For example, if you drifted in a canoe from Samoa you would have only one chance in 700 of ending up in the Marquesas. Yet we know that people did make such journeys across the Pacific in an easterly direction. The obvious conclusion was that the voyaging was not accidental but intentional: the seafarers must have deliberately sailed against the wind and currents to get to the places where they eventually settled.

This computer simulation and David Lewis's work were compounded in 1976, when a replica Polynesian voyaging canoe was built in Hawai'i. The *Hokule'a* was a double-hulled canoe modelled on ancient Polynesian designs, but made in fibreglass rather than timber. After it was sailed for several years around the Hawai'ian islands, a voyage from Hawai'i to Tahiti was planned. Traditional Polynesian navigational techniques would be used, with Mau Piailug and David Lewis as crew members.

The voyage of around 3,000 kilometres lasted just under a month. The crew sailed without maps, compasses or sextant. The only means of navigation was Mau Piailug. The Micronesian had never sailed the route before, but he knew from knowledge handed down to him by his grandfather the route that should be taken to make landfall in Tahiti. On the journey they were followed by a *National Geographic* support boat, so Piailug's route was able to be scientifically checked. It turned out that his daily calculations about where they were in the Pacific Ocean were never more than 65 kilometres out. When you think of all the forces acting on the canoe—the different wind patterns, the fact the night sky was often clouded over so Piailug couldn't see the star positions and had to rely on currents and swells—it was an astounding feat. When the canoe arrived in Tahiti, the Tahitians declared a national holiday to celebrate the reawakening of the lost art of navigating.

The *Hokule'a*, a double-hulled canoe which replicates early
Pacific seafaring craft, completed an astonishing voyage
from Hawai'i to Tahiti in 1976, using only early Polynesian
navigational methods. Since then it has completed a further
seven Pacific voyages. Monte Costa Photography

Since the construction of the *Hokule'a*, a number of other ethnographic replica canoes have been built in the Pacific. In Northland, New Zealand, a man called Hek Busby made *Te Aurere*, a double-hulled ocean voyaging canoe that has since sailed around the Pacific a number of times. Another New Zealander, Greg Whakataka-Brightwell from Ngati Toa on the east coast, built a canoe, *Hawaiki-nui*, in Tahiti of New Zealand totara. Sir Tom Davis, the former premier of the Cook Islands, built a double-hulled voyaging canoe, *Te Au-o-Tonga*, and formed the Cook Island Voyaging Society, with plans to sail to South America and retrace the voyage where Polynesians picked up the kumara and brought them back into the Pacific. All these canoes have been involved in retracing the steps of our ancestors, and reinvigorating Polynesians' awareness of their maritime legacy.

Today, then, we know essentially where the Polynesians came from and we know roughly when. We know they sailed the biggest ocean in the world relatively easily: the Pacific must have been virtually buzzing with canoes at some time in the ancient past. The big question remains why. Why did Polynesians make these amazing voyages into the Pacific? And why did they eventually settle all of the islands of Polynesia that are capable of sustaining human life?

One theory centres on wanderlust: Polynesians simply wanted to see what was over the horizon so they went and had a look. This idea of an innate curiosity and a tremendous desire to explore is bound to be right at one level. There is no doubt that early Polynesians had the technical capabilities to sail easily to virtually anywhere they wanted, so what was to stop them going and having a look and maybe finding a new island to live on? That must have been part of that story.

Another associated theory is that of overpopulation: Polynesian communities on small tropical islands quickly grew to a point where pressure on resources was too great, and people were forced to leave and find new homelands. Presumably, they travelled for the most part in an easterly direction, although we do know that some went west again, back into Melanesia, to occupy the islands we refer to today as the Polynesian outliers.

Demographers, however, tell us that people appear to have left for new islands before population levels on their home islands reached critical levels. If the demographers are right, over-population may not be the critical factor. It is bound to be part of the story but it can't be the only explanation.

Another very popular theory, sometimes expressed in the anthropological literature, is that migration was spurred by sibling rivalry—fighting between brothers and sisters. Polynesian societies are very hierarchical. At the top of the social pyramid is the chief, and at the bottom are the commoners and slaves. The various layers in between depend on which Polynesian society and culture you're talking about, but the whole structure is inherently un-stable. If you take a society like that and place it on a small tropical island where the population is going to grow rapidly, some anthro-pologists argue that there is only one possible outcome and that is tension.

This is supported by a large number of stories in Polynesian mythology in which one way out for people stuck in the middle of the social order but politically ambitious is to build a canoe with your family, head off out to sea, find a new island, colonise it and become the new chief. If you're the younger brother of the chief, there is another alternative: you kill your older brother and assume his position. Both sets of stories are found. The truth, though, is we don't know why Polynesia was settled the way it was. There are big gaps in our knowledge. ...

We know, of course, that New Zealand was settled by people from East Polynesia. There are arguments about precisely when this happened, but probably it was around about 1200 to 1300 AD. Because of the ocean currents, you can't get to New Zealand accidentally from East Polynesia: you have to be sailing in this direction quite deliberately. Semantically, you can call New Zealand's discovery an 'accident' because people did not know it was here in the first place. But it was clearly found by someone sailing this way deliberately. And chances are that whoever that person was—Kupe, Toi or someone else—they probably turned around, sailed home and told all their relatives about this wonderful new land in the South Pacific, and that started a series of migrations.

The best we can say at the moment is that these people came from East Polynesia. We can't hone in on any particular island homeland—mainly because the material culture, the artifacts, of that time from the various islands of East Polynesia are so similar it is impossible to distinguish between them. But in the not-too-distant future, through the study of human DNA, we will undoubtedly learn a lot more about those migrations. We will know precisely who the nearest relatives, genetically speaking, are of Maori people.

From Peter Adds, 'A Brilliant Civilisation' in *The Transit of Venus*: Awa Press, Wellington, 2007.

The Transit of Venus

In 1642, Dutch mariner Abel Tasman, on an exploratory voyage to the South Pacific, became the first European to sight the land where Maori had settled. The first European scientists to visit New Zealand—the botanists Joseph Banks and Daniel Solander on board the *Endeavour* captained by James Cook—came a century later. The main purpose of Cook's voyage, which was sponsored by the Royal Society, was to observe the 1769 transit of Venus—the passage of the planet Venus across the Sun—from Tahiti. This was part of a vast international effort, and one of the earliest global scientific collaborations. By comparing observations of the transit from different locations on Earth, scientists hoped to calculate the distance between the Sun and the Earth.

Transits of Venus take place in pairs separated by eight years, and these happen only every 105 to 121 years. Chris Orsman wrote this poem for the transit that took place on June 8, 2004.

Transit

An eye in the curve of a bay,
a pupil in the circle of a lens,
an eyelash sweeping the glass
clean of the smallest grit
on the surface of the mind,
so that, way out there,
Venus becomes a barque
moving across its measurable
sea of light, shadowing forth
the noble angle of the parallax.
Passage of island to island,
a leather-clad observatory
hoisted into a tree in a remote cove.

The quest to fix exact longitude
hangs on a held-in breath,
an eye on a marine clock,
as the planet's speck touches the limb
and the count-down begins.
Our place is fixed among the stars,
but not among the continents
which drift on currents of their own,
tectonic plates carrying us to other,
more imagined destinations.
And something is always occluding
our certainties, pulling the perfect
sphere of our knowledge out of shape,
like Venus on the Sun's edge,
turning us back to our equations,
with the tides lapping the sand
or caulked boards beneath our feet.
So we come to this apt conclusion:
that here is as good as anywhere
to value the constancy of the planets
in transit, the calculations
made in companionable silence,
that turn out slightly wrong,
dangers coolly perceived and faced,
shoals and reefs that remind us
not to get too comfortable with life,
to relish the dangerous edge
of the Universe—the trajectory rising
to a sudden flare-up in the heavens
that's not a comet or a meteor
but a bright cargo of human hopes
greeting the empyrean.

From *2004 Annual Report of the Royal Society of New Zealand*:
www.rsnz.org/directory/yearbooks/2004.

General Account
of New Zealand

Following his 1769 observation of the transit of Venus from Tahiti, Captain James Cook opened a sealed set of instructions from the Royal Society. They directed him to sail south in search of Terra Australis Incognita, the unknown southern continent. The large continent the society expected was not there, but on October 6, after they had sailed west, Nicholas Young, the twelve-year-old assistant to the ship's surgeon, spied a headland of the east coast of the North Island of New Zealand. Two days later Cook made landfall at Poverty Bay. He and his crew then spent six months circumnavigating New Zealand and exploring the coastal regions.

The ship's botanist Joseph Banks (1743–1820) wrote this summary account of New Zealand's landscape, flora and fauna, focusing on these features as resources ripe for exploitation. Banks, a wealthy young man, was travelling with an extensive natural history library, a team of artists and servants, and the respected Swedish botanist Daniel Solander.

The face of the country is in general mountainous, especially inland, where probably runs a chain of very high hills, parts of which we saw at several times. They were generally covered with snow, and certainly very high…

The sea coast, should it ever be examined, will probably be found to abound in good harbours. We saw several, of which the Bay of Islands, or Motuaro, and Queen Charlotte's Sound, or Totarra-nue, are as good as any which seamen need desire to come into, either for good anchorage or for convenience of wooding and watering. The outer ridge of land which is open to the sea is (as I believe is the case of most countries) generally barren, especially to the southward, but within that the hills are covered with thick woods quite to the top, and every valley produces a rivulet of water.

The soil is in general light, and consequently admirably adapted to the uses for which the natives cultivate it, their crops consisting entirely of roots. On the southern and western sides it is the most barren, the sea being generally bounded either by steep hills or vast tracts of sand, which is probably the reason why the people in these parts were so much less numerous, and lived almost entirely upon fish. The northern and eastern shores make, however, some amends for the barrenness of the others; on them we often saw very large tracts of ground, which either actually were, or very lately had been, cultivated, and immense areas of woodland which were yet uncleared, but promised great returns to the people who would take the trouble of clearing them. Taoneroa, or Poverty Bay, and Tolago [Tolaga] especially, besides swamps which might doubtless easily be drained, sufficiently evinced the richness of their soil by the great size of all the plants that grew upon them, and more especially of the timber trees, which were the straightest, cleanest, and I may say the largest I have ever seen, at least speaking of them in the gross. I may have seen several times single trees larger than any I observed among these; but it was not one, but all these trees, which were enormous, and doubtless had we had time and opportunity to search, we might have found larger ones than any we saw, as we were never but once ashore among them, and that only for a short time on the banks of the river Thames, where we rowed for many miles between woods of these trees, to which we could see no bounds. The river Thames is indeed, in every respect, the most proper place we have yet seen for establishing a colony. A ship as large as ours might be carried several miles up the river, where she could be moored to the trees as safely as alongside a wharf in London river, a safe and sure retreat in case of an attack from the natives. Or she might even be laid on the mud and a bridge built to her. The noble timber of which there is such abundance would furnish plenty of materials for building either defences, houses, or vessels; the river would furnish plenty of fish, and the soil make ample returns for any European vegetables etc. sown in it.

I have some reason to think from observations made upon the vegetables that the winters here are extremely mild, much more so than in England; the summers we have found to be scarcely at all hotter, though more equally warm.

The southern part, which is much more hilly and barren than the northern, I firmly believe to abound with minerals in a very high degree: this, however, is only conjecture. I had not to my great regret an opportunity of landing in any place where the signs of them were promising, except the last; nor indeed in any one, where from the ship the country appeared likely to produce them, which it did to the southward in a very high degree, as I have mentioned in my daily Journal.

On every occasion when we landed in this country, we have seen, I had almost said, no quadrupeds originally natives of it. Dogs and rats, indeed, there are, the former as in other countries companions of the men, and the latter probably brought hither by the men; especially as they are so scarce, that I myself have not had opportunity of seeing even one. Of seals, indeed, we have seen a few, and one sea-lion; but these were in the sea, and are certainly very scarce, as there were no signs of them among the natives, except a few teeth of the latter, which they make into a kind of bodkin and value much. It appears not improbable that there really are no other species of quadrupeds in the country, for the natives, whose chief luxury in dress consists in the skins and hair of dogs and the skins of diverse birds, and who wear for

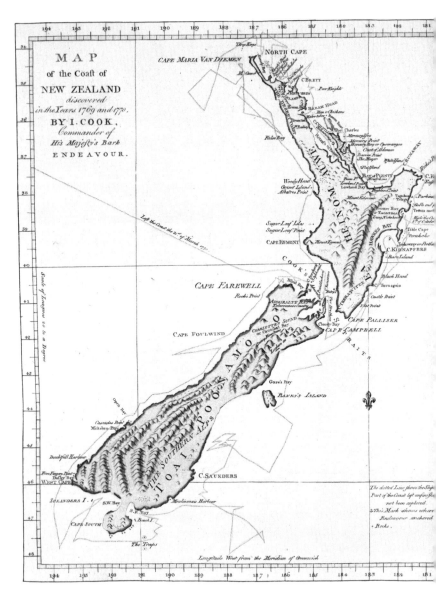

James Cook's map of New Zealand, showing the path of the
Endeavour's 1769–70 circumnavigation of New Zealand.
PUBL 0037-25, Alexander Turnbull Library, Wellington

ornaments the bones and beaks of birds and teeth of dogs, would probably have made use of some part of any other animal they were acquainted with, a circumstance which, though carefully sought after, we never saw the least signs of.

Of birds there are not many species, and none, except perhaps the gannet, are the same as those of Europe. There are ducks and shags of several kinds, sufficiently like the European ones to be called the same by the seamen, both which we ate and accounted good food, especially the former, which are not at all inferior to those of Europe.

Besides these there are hawks, owls, and quails, differing but little at first sight from those of Europe, and several small birds that sing much more melodiously than any I have heard. The sea coast is also frequently visited by many oceanic birds, as albatrosses, shearwaters, pintados etc., and has also a few of the birds called by Sir John Narbrough penguins, which are truly what the French call a *nuance* between birds and fishes, as their feathers, especially on their wings, differ but little from scales; and their wings themselves, which they use only in diving, by no means attempting to fly or even accelerate their motion on the surface of the water (as young birds are observed to do), might thence almost as properly be called fins.

Neither are insects in greater plenty than birds; a few butterflies and beetles, flesh-flies very like those in Europe, mosquitos and sand-flies, perhaps exactly the same as those of North America, make up the whole list. Of these last, however, which are most justly accounted the curse of any country where they abound, we never met with any great abundance; a few indeed there were in almost every place we went into, but never enough to make any occupations ashore troublesome, or to give occasion for using shades for the face, which we had brought out to protect us from such insects.

For this scarcity of animals on the land the sea, however, makes abundant recompense; every creek and corner produces abundance of fish, not only wholesome, but at least as well-tasted as our fish in Europe. The ship seldom anchored in, or indeed passed over (in light winds), any place whose bottom was such as fish generally resort to, without our catching as many with hooks and line as the people could eat. This was especially the case to the southward, where, when we lay at anchor, the boats could take any quantity near the rocks; besides which the seine seldom failed of success, insomuch that on the two occasions when we anchored to the southward of Cook's Straits, every mess in the ship that had prudence enough salted as much fish as lasted them many weeks after they went to sea.

For the sorts, there are mackerel of several kinds, one precisely the same as our English, and another much like our horse-mackeral, besides several more. These come in immense shoals and are taken in large seines by the natives, from whom we bought them at very easy rates. Besides these there were many species which, though they did not at all resemble any fish that I at least have before seen, our seamen contrived to give names to, so that hake, bream, cole-fish etc. were appellations familiar with us, and I must say that those which bear these names in England need not be ashamed of their namesakes in this country. But above all the luxuries we met with, the lobsters, or sea-crawfish, must not be forgotten. They are possibly the same as are mentioned in Lord Anson's voyage as being found at the island of Juan Fernandez, and differ from ours in England in having many more prickles on their backs and being red when taken out of the water. Of them we bought great quantities everywhere to the northward from the natives, who catch them by diving near the shore, feeling first with their feet till they find out where they lie. We had also that fish described by Frézier in his voyage to Spanish South America by the name of *elefant, pejegallo,* or *poisson coq,* which, though coarse, we made shift to eat, and several species of skate or stingrays, which were abominably coarse. But to make amends for that, we had among several sorts of dogfish one that was spotted with a few white spots, whose flavour was similar to, but much more delicate than, our skate. We had flat fish also like soles and flounders, eels and congers of several sorts, and many others, which any European who may come here after us will not fail to find the advantage of, besides excellent oysters, cockles, clams, and many other sorts of shellfish etc.

Though the country generally is covered with an abundant verdure of grass and trees, yet I cannot say that it is productive of such great variety as many countries I have seen: the entire novelty, however, of the greater part of what we found recompensed us as natural historians for the want of variety. Sow-thistle, garden-nightshade, and perhaps one or two kinds of grasses, were exactly the same as in England, three or four kinds of fern were the same as those of the West Indies: these with a plant or two common to all the world, were all that had been described by any botanist out of about four hundred species, except five or six which we ourselves had before seen in Tierra del Fuego.

Of eatable vegetables there are very few; we, indeed, as people who had been long at sea, found great benefit in the article of health by eating plentifully of wild celery and a kind of cress which grows everywhere abundantly near the sea-side. We also once or twice met with a herb like that which the country people in England call 'lamb's-quarters' or 'fat-hen,' which we boiled instead of greens; and once only a cabbage-tree, the cabbage of which made us one delicious meal. These, with the fern roots and one vegetable (*Pandanus*) totally unknown in Europe, which, though eaten by the natives, no European will probably ever relish, are the whole of the vegetables which I know to be eatable, except those

An illustration by Sydney Parkinson (1745?–71), a Scottish natural history artist whom Joseph Banks employed to accompany him to the Pacific, of *Metrosideros tomentosa*, or pohutukawa, 'which bears a very conspicuous scarlet flower made up of many threads, and which is as big as an oak in England'.

B-026-023, Alexander Turnbull Library, Wellington

which they cultivate and have probably brought with them from the country from whence they themselves originally come.

Nor does their cultivated ground produce many species of esculent plants; three only have I seen, yams, sweet potatoes, and cocos, all three well known and much esteemed in both the East and West Indies. Of these, especially the two former, they cultivate often patches of many acres, and I believe that any ship that found itself to the northward in the autumn, about the time of digging them up, might purchase any quantity. They also cultivate gourds, the fruits of which serve to make bottles, jugs etc., and a very small quantity of the Chinese paper mulberry tree.

Fruits they have none, except I should reckon a few kinds of insipid berries which had neither sweetness nor flavour to recommend them, and which none but the boys took the pains to gather.

The woods, however, abound in excellent timber, fit for any kind of building in size, grain, and apparent durability. One, which bears a very conspicuous scarlet flower made up of many threads, and which is as big as an oak in England, has a very heavy hard wood which seems well adapted for the cogs of mill-wheels etc., or any purpose for which very hard wood is used. That which I have before mentioned to grow in the swamps, which has a leaf not unlike a yew and bears small bunches of berries, is tall, straight, and thick enough to make masts for vessels of any size, and seems likewise by the straight direction of the fibres to be tough, but it is too heavy. This, however, I have been told, is the case with the pitch-pine in North America, the timber of which this much resembles, and which the North Americans lighten by tapping, and actually use for masts.

But of all the plants we have seen among these people, that which is the most excellent in its kind, and which really excels most if not all that are put to the same uses in other countries, is the plant which serves them instead of hemp or flax. Of this there are two sorts. The leaves of both much resemble those of flags; the flowers are smaller and grow many more together. In one sort they are yellowish, in the other of a deep red. Of the leaves of these plants all their common wearing apparel is made with very little preparation, and all strings, lines, and cordage for every purpose, and that of a strength so much superior to hemp as scarce to bear comparison with it. From these leaves also by another preparation a kind of snow-white fibre is drawn, shining almost as silk, and likewise surprisingly strong; of this all their finer cloths are made: their fishing-nets are also made of these leaves, without any other preparation than splitting them into proper breadths and tying the strips together. So useful a plant would doubtless be a great acquisition to England, especially as one might hope it would thrive there with little trouble, as it seems hardy and affects no particular soil, being found equally on hills and in valleys, in dry soil and the deepest bogs, which last land it seems, however, rather to prefer, as I have always seen it in such places of a larger size than anywhere else.

From *Journal of the Right Hon. Sir Joseph Banks*, edited by
Sir Joseph D. Hooker: Macmillan and Co. Ltd, London, 1896.

'Not a Pleasant Place'

The English naturalist Charles Darwin (1809–82) visited New Zealand in 1835, near the end of his five-year voyage on the HMS *Beagle,* during which he began to form the ideas that would be expressed in his 1859 book *On the Origin of Species by Means of Natural Selection.* It is exciting to wonder what Darwin made of the country's primitive flora, giant carnivorous land snails and large flightless birds—described by the twentieth-century naturalist George Gibbs as 'outlandish freaks' of the natural world. Sadly, though, Darwin gave little indication of having noticed the native fauna and flora, taking more joy in the plums and potatoes growing in the neatly planted gardens of the English missionaries.

Darwin wrote this account in his *Beagle* diaries before he had fully formed his ideas about natural selection. However, it is already clear that he has ideas about superior and inferior species (and races). His later work on evolution, with its concept of the survival of the fittest, would be used to support the notion that New Zealand's native species were in decline, doomed to be replaced by 'superior' or 'fitter' species imported from Europe.

The *Beagle* arrived in the Bay of Islands on December 21, and after nine days it left for Sydney, a place Darwin found much more to his liking.

*December 23rd…*We now commenced our walk; the road lay along a well beaten path, bordered on each side by the tall fern which covers the whole country. After travelling some miles, we came to a little country village, where a few hovels were collected together and some patches of ground cultivated for potato crops. The introduction of the potato has been of the most essential benefit to the island; it is now much more used than any native vegetable. New Zealand is favoured by one great natural advantage, namely, that the inhabitants can never perish from famine. The whole country abounds with fern, and the roots of this, if not very palatable, yet contain much nutriment. A native can always subsist on them and on the shellfish, which is very abundant on all parts of the sea shore. The villages are chiefly conspicuous by the platforms which are raised on four posts, ten or twelve feet above the ground and on which the produce of the fields is kept secure from all accidents. On coming near to one of the huts, I was much amused by seeing in due form the ceremony of rubbing, or as it would be more properly called, pressing noses...

The ceremony of pressing noses having been completed with all present, we seated ourselves in a circle in the front of one of the houses and rested there half an hour. All the native hovels which I have seen, have nearly the same form and dimensions and all agree in being filthily dirty. They resemble a cow shed with one end open; but having a partition a little way within, with a square hole in it, which cuts off a part and makes a small gloomy chamber. When the weather is cold the inhabitants sleep there and likewise keep all their property. They eat, however, and pass their time in the open part in front.

My guides having finished their pipes, we continued our walk. The path led through the same undulating country, the whole uniformly clothed as before with fern. On our right hand we had a serpentine river, the banks of which were fringed with trees and here and there on the hill sides there were clumps of wood. The whole scene, in spite of its green colour, bore rather a desolate aspect. The sight of so much fern impresses the mind with an idea of useless sterility; this, however, is not the case, for wherever the fern grows thick and breast high, the land by tillage becomes productive. I have heard it asserted, and I think with much probability, that all this extensive open country was once covered by forests, and that it had been cleared ages past by the aid of fire.

It is said that frequently by digging in the barest spots, lumps of that kind of resin, which flows from the kauri pine, are found. The natives had an evident motive in thus clearing the country, for in such parts the fern, formerly so staple an article of food, best flourishes. The almost entire absence of associated grasses which forms so remarkable a feature in the vegetation of this Island, may perhaps be accounted for, by the open parts being the work of man, while Nature had designed the country for forest land. The soil is volcanic, in several parts we passed over slaggy and vesicular lavas, and the form of a crater was clearly to be distinguished in several of the neighbouring hills. Although the scenery is nowhere beautiful, and only occasionally pretty, I enjoyed my walk…

At length we reached Waimate; after having passed over so many miles of an uninhabited useless country, the sudden appearance of an English farm house and its well dressed fields, placed there as if by an enchanter's wand, was exceedingly pleasing. Mr Williams not being at home, I received in Mr Davies' house a cordial and pleasant welcome. After drinking tea with his family party, we took a stroll about the farm. At Waimate there are three large houses, where the missionary gentlemen, Messrs Williams, Davies and Clarke, reside; near to these are the huts of the native labourers. On an adjoining slope fine crops of barley and wheat in full ear, and others of potatoes and of clover, were standing; but I cannot attempt to describe all I saw; there were large gardens, with every fruit and vegetable which England produces, and many belonging to a warmer clime. I may instance asparagus, kidney beans, cucumbers, rhubarb, apples and pears, figs, peaches, apricots, grapes, olives, gooseberries, currants, hops, gorse for fences, and English oaks! and many different kinds of flowers. Around the farmyard were stables, a threshing barn with its winnowing machine, a blacksmith's forge, and on the ground, ploughshares and other tools; in the middle was that happy mixture of pigs and poultry which may be seen so comfortably lying together in every English farmyard. At the distance of a few hundred yards, where the water of a little rill has been dammed up into a pool, a large and substantial water-mill had been erected. All this is very surprising when it is considered that five years ago, nothing but the fern here flourished. Moreover native workmanship, taught by the missionaries, has effected this change—the lesson of the missionary is the enchanter's wand. The house has been built, the windows framed, the fields ploughed, even the trees grafted by the New Zealander. At the mill a New Zealander may be seen powdered white with flour, like his brother miller in England. When I looked at this whole scene I thought it admirable. It was not that England was vividly brought before my mind; yet as the evening drew to a close, the domestic sounds, the fields of corn, the distant country, with its trees, now appearing like pasture land, all might well be mistaken for such. Nor was it the triumphant feeling at seeing what Englishmen could effect: but a thing of far more consequence; the object for which this labour had been bestowed —the moral effect on the native inhabitant of New Zealand. ...

December 24th. In the morning prayers were read in the native tongue to the whole family: after breakfast I rambled about the gardens and farm. This was market day when the natives of the surrounding hamlets bring their stock of potatoes, Indian corn or pigs, to exchange for blankets, tobacco and sometimes (from the persuasions of the missionaries) for soap. Mr Davies' eldest son, who manages a farm of his own, is the man of business in the market. The children of the missionaries who came whilst young to the Island, understood the language better than their parents, and can get anything more easily done by the natives. Mr Williams and Mr Davies walked with me to part of a neighbouring forest to show me the famous kauri pine. I measured one of these noble trees and found it to be thirty-one feet in circumference; there was another close by which I did not see, thirty-three, and I have heard of one no less than forty feet. The trunks are also very remarkable by their smoothness, cylindrical figure, absence of branches, and having nearly the same girth for a length from sixty even to ninety feet. The crown of this tree where it is irregularly branched is small and out of proportion to the trunk; and the foliage is again diminutive as compared to the branches. The forest in this part was almost composed of the kauri; amongst which the great ones from the parallelism of their sides stood up like gigantic columns of wood. The timber of this tree is the most valuable product of the island; besides this, quantities of a resin oozes from the bark, which is collected and sold at a penny a pound to the North Americans, but its use is kept secret. On the outskirts of the wood I saw plenty of the New Zealand hemp plant growing in the swamps; this is the second most valuable export. This plant resembles (but not botanically) the common iris: the under surface of the leaf is lined by a layer of strong silky fibres; the upper green vegetable matter being scraped off with a broken shell, the hemp remains in the hand of the workwoman. In the forest besides the kauri there are some fine timber trees: I saw numbers of beautiful tree-ferns and heard of palms. Some of the New Zealand forests must be im-penetrable to a very extraordinary degree; Mr Matthews gave me an account of one which although only thirty-four miles in width and separating two inhabited districts, like the central forest of Chiloé,* had never been passed. He and another missionary each with a party of about fifty men, undertook to open a road; but it cost them more than a fortnight's labour! In the woods I saw very few birds; with respect to animals it is very remarkable that so large an island, extending over nearly a thousand miles in latitude, and

---

* Chiloé Island, off the coast of southern Chile.

in many parts one hundred and fifty broad, with varied stations, a fine climate and land of all heights from 14,000 feet downwards, should not possess one indigenous animal with the exception of a small rat. It is moreover said that the introduction of the common Norway kind has entirely annihilated the New Zealand species in the short space of two years, from the northern extremity of the island. In many places I noticed several sorts of weeds which, like the rats, I was forced to own as countrymen. A leek, however, which has overrun whole districts and will be very troublesome, was imported lately as a favour by a French vessel. The common dock is widely disseminated and will I am afraid for ever remain a proof of the rascality of an Englishman who sold the seeds for those of the tobacco plant.

On returning from our pleasant walk to the houses, I dined with Mr Williams; and then a horse being lent me, I returned to the Bay of Islands. I took leave of the missionaries, with thankfulness for their kind welcome and with feelings of high respect for their gentleman-like, useful and upright characters. I think it would be difficult to find a body of men better adapted for the high office which they fulfil. ...

December 27th–29th. Chiefly employed in writing letters, and in collecting some specimens. (*30th.*) In the afternoon we stood out of the Bay of Islands on our course to Sydney. I believe we were all glad to leave New Zealand; it is not a pleasant place; amongst the natives there is absent that charming simplicity which is found at Tahiti; and of the English the greater part are the very refuse of society. Neither is the country itself attractive...

From *Charles Darwin's Diary of the Voyage of H.M.S. 'Beagle'*, edited by Nora Barlow: Cambridge University Press, Cambridge, 1933.

Whales and Whaling

Ernst Dieffenbach (1811–55) was the first trained scientist to live and work in New Zealand. In 1840, the German scientist sailed to Wellington on the *Tory* as surgeon and naturalist for the New Zealand Company. His task was to report on the country's flora, fauna and mineral and water resources so the company could assess the suitability of different areas for settlement. After returning to London, Dieffenbach published the two-volume work *Travels in New Zealand*, from which this account of nineteenth-century whales and whaling comes. Dieffenbach's warning that the whale fishery would collapse under the system of shore whaling was prescient: by the 1850s the fishery was in decline, and by the end of the nineteenth century right whales were rarely seen in New Zealand waters.

Almost all the whales which are killed on the shores of New Zealand are females, or cows, and their calves. The male, or bull, is very rarely caught, as it never approaches the land so near as the female, and is more shy and wild. The season in which whaling is carried on is from May to October. In the beginning of May the cows approach the shallow coasts and smooth waters for the purpose of bringing forth their young. This period lasts about four months, as in May whales are seen with newly born calves, and cows have been killed in July in full gestation. During the same months also copulation is sometimes observed by the whalers. But from these data it is impossible to draw a conclusion on the real period of gestation in these huge animals, which has never yet been satisfactorily determined. In company with the cows are also the calves of the preceding year or years, for it is still uncertain at what age the whale attains its full size and leaves the mother; these young whales are called scrags, and they yield about four tuns of oil.[*]

It appears that the female generally produces but one calf at a birth: the cow is indeed sometimes seen with two; and although in this case it is the opinion of the whalers that one is an orphan calf, yet it is probable that the black whale, like the northern sperm whale, occasionally produces twins. A calf, which appeared full grown, and which was cut out of the mother a short time before my arrival in Te-awa-iti,[†] measured fourteen feet.

The whale is a truly migratory animal, and its migrations are the most interesting part of its history. They arrive at the coasts of New Zealand in the beginning of May from the northward, go through Cook's Straits, keeping along the coast of the northern island, and pass between the latter and Entry Island.[††] This is borne out by the fact that they are never seen on the opposite coast, nor do they enter the northern entrance of Queen Charlotte's Sound. From Entry Island they sweep into Cloudy Bay, and at the end of October they go either to the eastward or return to the northward. In the beginning of the season the chase is said to be most successful in Cook's Straits and Te-awa-iti; in the three latter months in Port Underwood, which is only thirty miles distant. From the month of June they begin to show themselves near the Chatham Islands, 150 leagues to the eastward of New Zealand, where their number increases with the termination of the season in the latter place. During the six remaining months of the year the ships cruising in the 'whaling ground' fall in with many whales. This whaling ground extends from the Chatham Islands to the eastward of the northern island of New Zealand, and from thence to Norfolk Island. It is curious that the whalers assert that this

---

[*] A tun of whale oil was 955 litres.
[†] A large whaling settlement on Arapawa Island in Tory Channel.
[††] Kapiti Island.

whaling ground is nothing but a shoal, although I am not aware that soundings have ever been obtained. Perhaps Captain Ross, who is now in the South Seas provided with sounding-lines, will confirm a fact which is of some importance in the natural history of this animal. The migration of the whale is probably owing to its search for food; but we must still regard it as a subject for inquiry, which cannot be terminated before we know many more particulars connected with it, and especially how far it depends on the greater or lesser quantity, in certain localities and at certain seasons, of the small animal of the medusa kind upon which the black whales feed. Their approach to the shores of New Zealand, however, is particularly connected with the process of parturition, as I have already mentioned. In the month of June they are observed in the same condition, viz. with calves, at the Cape of Good Hope. It seems as if certain herds of whales, if I may be allowed to use that term, which occupy a limited district, visit at the end of the period of gestation the bights and inlets of those countries which are next to their feeding-grounds: the same is the case round Van Diemen's Land. But it has yet to be proved that the black whale of the Cape of Good Hope is the same with the black whale of New Zealand.

Besides this general migration, which, until more accurate data are obtained, I do not conceive should be termed a circumnavigation, but merely a migration of different species in a certain marine district, there exists also a daily one. These fish approach the shores and bays with the flood-tide, and quit them with the ebb. In their general migrations, also, they seem to be influenced by the direction of the tides. Whales are often seen in places where the depth of water does not much exceed their own breadth, rubbing their huge bodies against the rocks, and freeing themselves of the barnacles and other parasitical animals with which they are covered.

The maternal affection of the whale for its young is very great. As soon as the mother observes a threatened danger, she clings, as it were, to the calf, tries to hide it, and often takes it between her flooks (fins) and attempts to escape. She has even been observed to carry off the calf when it had been killed, but not fastened upon. Sometimes, however, she seems to be infatuated, and heedless of all that passes around her. If the calf has been once fastened upon, the mother will never leave it. The whalers assert that the young cows have less affection for their offspring than the old ones, and will desert them at the appearance of the least danger. It is, however, the affection of the whale for her young which becomes the principal means of her destruction. The calf, inexperienced and slow, is easily killed, and the cow is afterwards a sure prey.

It is not known in what position the cow suckles her calf. The teats, which are two in number, are abdominal, and situated between membranaceous folds on both sides of the genital organs. I was astonished to find them so small. In a female, whose mammae were full of a fat milk resembling cow-milk in taste, the teats were not larger than those of a cow. The operation of suckling never having been observed, it is no matter of surprise to find the whalers denying that the cows suckle their calves; there can, however, be no doubt of the fact.

The manner of carrying on whaling is so well known as to render it unnecessary for me to dwell upon it at any length. The whale-boats are admirably adapted for the purpose for which they are intended. They are of various construction, and are designated as English, French, or American: each has some peculiarity to recommend it. They are capable of resisting the rough sea of Cook's Straits, but are at the same time swift and buoyant. When starting on a whaling expedition, the boats leave Te-awa-iti before the dawn of the morning. Each has either five or six oars, and a crew accordingly. The boat-steerer and headsmen are the principal men in the boat, and are generally Europeans; the rest are natives. They pull to the entrance of Tory Channel, where a view opens over Cook's Straits and Cloudy Bay from the southern headland, where they keep a 'look-out' for the spouting of a whale. The boat which kills the calf claims the cow, even though it should have been killed by another boat's crew. If a whale has been killed, the different boats assist each other in towing it to Te-awa-iti. I once saw ten or twelve boats towing-in a whale. Each boat had a little flag, and the whole scene was gay and animated. One day a calf had been killed, and the cow, having been fastened upon, but not despatched, was towed inside the channel. Gasping in the agonies of death, the tortured animal, when close to our ship, threw up jets of blood, which dyed the sea all around; and, beating about with its tail, it broke a boat right in the middle, and threw the crew into the water; but it at length died, exhausted from the many wounds which the irons and harpoons had inflicted. The calf was stated by the whalers to be six weeks old (on what grounds I do not know), and was twenty-four feet long. It was cut up in a few minutes, and gave several barrels of oil. The process was so rapid, that when I came ashore I found only the head. I cut out the brains, the weight of which, amounting to five pounds and one ounce, astonished me greatly. The whalebone was very soft, and therefore useless. There were two hundred plates of it on each side of the roof of the upper jaw. I got the whole roof cut off, and, intending to dry and preserve it, I placed it on the roof of a native house; but on the following morning I had the mortification to find that the rats and native dogs had found their way to it in the night, and had eaten all the softer parts, so that the rest fell to pieces.

A portion of the heart of this calf was roasted and sent to our table. In taste I found it very like beef, but it was darker in colour. The cow was sixty feet long, and measured between the fins on the belly eighty-two inches. Her skin was a velvet-like black, with the exception of a milk-white spot round the navel. As regards the colour of the whale, I have been repeatedly assured that it is sometimes speckled, and that even perfect albinos, or cream-coloured ones, are seen, which must indeed be beautiful animals. The fat or blubber of this whale was nine inches thick, and yielded eight tuns and a half of oil. Whales have been known to yield twelve or thirteen tuns; but I have been told that so large a quantity is now very rarely obtained, from the great decrease of the whales. A whale which yields nine tuns is at present regarded as a very good one.

The tongue was of a white or ash colour, blackish towards the root. This organ gave several barrels of oil, and is a monopoly of the 'tonguer,' or 'cutter-in.' The latter operation is performed in Te-awa-iti near the shores, where by means of a windlass the whale is raised to the surface of the water under a scaffold called the 'shears.' The blubber is cut off in square pieces by means of a sharp spade; it is then carried to the shore, and immediately put into the trying-pots. The 'cutting-up' of a whale, *secundum artem*, is a process which requires great proficiency, like that of the skilful dissector, who separates the cutis, and with it at once all fat and cellular tissue, from the subjacent muscles. In the whale the blubber is to be regarded as the cutis, in the cellular structure of which the oily matter has been deposited. Shortly after the death of the fish the epidermis comes off in large pieces, looking like oiled and dried satin.

As soon as the process of cutting was over, the natives, who had come with their canoes from the Sound, cut off large pieces of the flesh, which they carried off to feast upon. They also fished in the evening for sharks, and a curious gelatinous fish, which fastened in numbers on the sunken carcase, and which is nearly related to the *Myxine glutinosa* of our latitudes. …

I must observe here that the sort of shore-whaling which I have just described is very detrimental to the whale fishery in general, and the number of whales has decreased from year to year. The female whale approaches the land merely for the purpose of bringing forth and rearing her young. Later in the year, when the calf has attained a certain size, the cows leave the immediate neighbourhood of the coast, and return to the 'whaling ground', where the males share with them the dangers resulting from the pursuit of man. Would it not, therefore, be advisable by legislative enactments to put an end to the whale fishery from the coasts, and to restrict it to a certain distance from shore, where it would have to be pursued in ships? To kill the calves in order to capture the mother, or to kill the latter in the time of gestation, is an un-profitable and cruel proceeding; but it carries with it its own punishment. In a few years this trade, of which, from the geo-graphical position of the 'whaling ground', New Zealand might have continued to be the centre, will be annihilated. Seals, which were plentiful in New Zealand, but were slaughtered in the same indiscriminate manner, have already entirely disappeared. The protection proposed would only be such as the governments of all civilised nations have long bestowed on their coast fisheries during certain seasons. Unfortunately there exists a belief that the female whale in a state of gestation, or immediately afterwards, yields the greatest quantity of oil; but I have reason to believe that this is entirely unfounded. If we may judge from analogy of fishes, birds, and the whole class of mammiferous animals, the assertion must be untrue; and this view is confirmed by the testimony of those who carry on the fishery from ships, that, instead of the whales being fatter during that time, the contrary is the case, and that the average result of the fishing on the whaling grounds exceeds that of the coast fisheries. I have also heard that very often cows have been brought into Te-awa-iti which were remarkably lean, and did not yield more than five or six tuns of oil. We must also expect the oil from whales in the period of gestation to be inferior in quality, from the great change which then is effected in all the solid and fluid elements of the body.

Whilst I am thus pleading the cause of the whale, I am well aware that the most effective mode of preserving the fishery would be to spare the cows and calves altogether, and to kill merely the bulls. The whalers can distinguish at a considerable distance a bull from a cow—the elevation near the spout-holes, called the top-knot, being much higher in the bulls, and this part is always above the water; but such an extensive protection is probably impracticable.

It would suffice if, during the winter season, or from May to October, all whaling was prohibited within a certain distance from the shores of New Zealand, and a man-of-war cutter kept to enforce obedience to this rule.

From *Travels in New Zealand: With Contributions to the Geography, Geology, Botany, and Natural History of That Country* by Ernst Dieffenbach, volume one: J. Murray, London, 1843.

To Ernst Dieffenbach

The Wellington poet Ian Wedde was inspired by the work of Ernst Dieffenbach, and composed this poem entirely from words and phrases in Dieffenbach's 1843 book *Travels in New Zealand*.

Once thrown up by the mysterious fires of the deep
They were soon again hid from our view, basalts,
Greenstones, trachyte, augitic rock, oxydated
Iron-clay, its summit a never-failing
Object of attraction when the morning or evening sun

Gilded its snowy summit with a rosy hue.
Cliffs of yellow clay embedded with the remains
Of trees still existing in the island, and a countless
Number of small streams, all the pebbles
Hard blue trap-rock. Here and there

Might be seen a majestic rimu pine, or rata,
Bearing crimson flowers. The country now
Began to rise a little, we scarcely ever
Obtained a view of the sun, at sunset a thick
Forest surrounded this place. Rats

Ran over us all night. Another
Bird is very common, and always screams—
Huei, huei, tierawak, tierawak. For the first
Time, covered with snow, but its summit hid
In the clouds. Vegetation had long ceased,

IAN WEDDE

A supernatural spirit breathes on him
In the evening breeze, which gives birth to the poetry
Of infant nations—the old tales of the Germans.
How little benefit has resulted to barbarous tribes
Gradually, as if acted upon by a slow poison.

From *The Commonplace Odes* by Ian Wedde: Auckland University Press, Auckland, 2001.

Magnitude Eight Plus

The most powerful earthquake ever recorded in New Zealand is the magnitude 8.2 Wairarapa quake of January 23, 1855. Caused by rupture of the Wairarapa fault, which runs from Cook Strait along the eastern edge of the Rimutaka Range, the violent earthquake lasted more than 50 seconds, shaking buildings to the ground and causing landslides to block roads. There was only one fatality in Wellington—Baron von Alzdorf was killed when a brick chimney in his hotel collapsed—but two people were killed in the Manawatu, and a group of Maori in the Wairarapa died when their whare collapsed. Wellington geologist Rodney Grapes (born 1945) captured local response to the earthquake in his book *Magnitude Eight Plus*.

On the morning of Sunday 21 January 1855, the HMS *Pandora* —a two-masted barquenteen of some 319 tons under the command of Byron Drury—arrived in Wellington. The journey from Dunedin had been plagued by bad weather and the ship's company were glad to have finally arrived in the sheltered, safe anchorage of Lambton Harbour after so much 'tossing and tumbling about'.

The view of Wellington from *Pandora*'s anchorage in six fathoms of water was picturesque with its backdrop of wooded hills and water frontage. The town, with a population of about 3,200, was rather scattered, extending more than five kilometres around the shore of Lambton Harbour. Running along the edge of the water was the principal street, Lambton Quay, nearly two-and-a-half kilometres long. A sea wall had been built along part of the quay and around Clay Point, where the sides of the steep hill had been cut away sufficiently to allow the construction of an unsealed carriageway in front of a row of small houses, shops and inns. Between Lambton Quay and the thickly wooded Karori Hills rising abruptly to the west lay the elevated ground of Thorndon Flat, divided into irregular streets, with detached houses and gardens—Government House and offices, a hospital, a Protestant church, and further inland the Thorndon Barracks and many other buildings. Perched here and there were neat-looking, villa-like, white, wooden cottages with agreeable lawns and gardens. The elevated flat was fronted by a steep cliff that extended around Pipitea Point. On the cliff stood Pipitea Pa and along the base of it ran Thorndon Quay, lined with numerous small wooden houses. Facing the central part of the harbour lay Te Aro Flat, the eastern half a bog, the western part containing houses extending a considerable distance inland—this area being the oldest part of the town. In the background stood the Mt Cook Barracks and near the shore was Te Aro (Taranaki) Pa. Although several wooden jetties projected into Lambton Harbour for 30 metres or so, there was as yet no good landing place. …

Skirting the western edge of the harbour was a 'tolerably good and level' road to the far-famed Hutt Valley. The valley foreshore, known as 'Pita-one', consisted of sandy hillocks and swampy ground, which gave way to a forest of giant rimu, pukatea, matai and tall tree-ferns. These, in turn, seemed to extend unbroken to the distant converging hills far up the valley. Traversing the valley was the Hutt River … The lower part of the valley was dotted with clearings and clusters of cottages with 'English' orchards. Some shops, two hotels, the wooden-stockaded Fort Richmond, a school and two churches were concentrated near a bridge over the river about three kilometres from its mouth. …

This was *Pandora*'s second visit to Wellington since arriving in August 1852 charged with the task of surveying the coastline and harbours of New Zealand … The admiralty task would be completed the following year, but for now, Commander Drury had returned to mark the fifteenth anniversary of the settlement before sailing on to Nelson.

Anniversary holidays began on Monday with festivities and sports. The day's highlight was a whale-boat race … Tuesday was the eagerly anticipated race day at Burnham Water in the centre of Miramar peninsula, where a racecourse had been formed amongst the sand-hills. From early in the morning, crowds of punters, owners and their horses began to gather from Wellington and 'up country'. The first race was followed by the 'Ladies Purse' and the 'Scurry Stakes' and all were well contested until it began to rain in the early afternoon, forcing most of the race-goers to leave. A strong northwesterly came on as they made their way back to town across the exposed sandy expanse of the Rongotai isthmus, which joins Miramar to the mainland, and over the Hataitai saddle to Te Aro Flat. Towards evening the wind dropped and by about seven o'clock the weather became calm for a short while until the northwesterly gale resumed.

A few seconds before 9.17 p.m., at a depth of 25 kilometres below Cook Strait, some 40 kilometres southwest of Wellington, a large section of the Earth's crust suddenly ruptured, releasing an energy pulse 1,000 times more powerful than the Hiroshima atomic bomb. The shock wave radiated outwards and upwards at a speed approaching six kilometres per second.

Lieutenant Morton Jones was up on deck when at precisely 17 minutes past nine the *Pandora* was jolted by a sudden and very severe vibration as if it 'was grating heavily and rapidly over a sandbank or perhaps resembling more closely the motion of a railway carriage'. Within the next minute the vibration increased in intensity and violence; the *Pandora* slewed broadside to the wind setting all the ship's bells ringing and causing panic below. To Jones the sensation was most extraordinary and although his first impression was that the ship had grounded he recalled that it was anchored in six fathoms of water. He realised that he was in the midst of an earthquake. Others quickly joined him on deck and, during the lull that followed the shock, they distinctly heard shouts and the crash of falling chimneys and houses in the town. It was clear from the lights passing rapidly to and fro and other signs of commotion that the earthquake had been felt severely on shore.

After returning from the races at Burnham Water, John Jolliffe, surgeon on the *Pandora*, had called on the D'Arcy family in their small, four-bedroom wooden house on Te Aro Flat. The children had gone to bed and Nelly and her sister were at Captain Paul's house 100 metres away. Jolliffe and Mrs D'Arcy were engaged in a game of dominoes while Captain D'Arcy was slumbering on the sofa. Suddenly, they heard a rumble like a carriage passing over a wooden pavement, followed immediately by a violent shock. With the cry, 'It's an earthquake,' Jolliffe and Mrs D'Arcy jumped up from the table but could hardly keep their feet because the ground was moving in waves which jerked the house rapidly to and fro. The chimney collapsed, bricks falling into the parlour and covering everything with dust. Pictures leapt from the walls; ornaments on the shelf were thrown into the centre of the room; chairs, tables and everything moveable fell or were tossed about. The decanter and tumbler on the sideboard crashed to the floor, along with bottles of wine in the cupboard, their contents mingling with the 'confusion of tossing about and smashing'. Desperately trying to get the door open against the force of the shaking, Jolliffe finally managed to wrench it open to be confronted by Captain Maybin from next door who rushed into the house like a 'madman with his hands thrown up' and yelling for everyone 'to get out of the house for God's sake!' Managing to evacuate everyone without injury the men proceeded to extract beds and blankets and in a short time made the children 'as snug as possible and warm' for they only had their nightdresses on and 'required much covering up'.

A little before nine o'clock, eager to know the outcome of the horse races at Burnham Water, Charles Carter had gone along to the Royal Hotel on Lambton Quay for news. He had been seated in the back parlour only a short time when all of a sudden the hotel began to move violently 'as if some great force were exerted at each of its ends to pull it rapidly and horizontally backwards and forwards'. Amidst the noise of breaking and jingling bottles and glasses, which were packed close together on the shelves of the bar, Carter and the other patrons made a speedy exit. Outside on the road he could feel the ground heaving and rolling and, while running towards his house on Wellington Terrace where his wife and child were, he was surprised to find his feet leaving the road and the road rising up to meet them halfway. He had only gone about 100 metres when a crashing noise made him look up to see the council chamber enveloped in a cloud of dust. The ground was still heaving and shaking and as the dust began to clear he was able to see that the two-storey building had compacted into one, with the upper storey sitting on the broken timbers and rubble of the lower.

Thomas Pilcher, who lived in Manners Street … suddenly felt the house begin to shake. He and his uncle rushed to prevent glasses from falling out of the cupboard but the shaking intensified and they both had the feeling that the house was about to collapse. In the front room, books flew out of the bookcase, bottles were thrown from the cupboard, and the chimney came down. In Pilcher's bedroom where his aunt was resting, everything fell from the shelf that circled the room and his desk 'hit aunt in the neck in its descent'. In the back parlour, the floor quickly became covered with a mixture of glasses, mustard, brandy, sugar, salt, pepper and broken bottles, all 'as if for a pie'.

Fortunately, the clock which tumbled from the mantelpiece did not break, but the chimney fell, smashing three rafters in its descent. In the kitchen, plates, glass bottles, kettles, water and everything else on the dresser were 'pitched to the ground'. Fortunately, the iron chimney survived, but all the brickwork around the stove was cracked. In the wash-house the chimney fell and other brickwork cracked; in the bakehouse both ovens were shattered as well as the chimney. The mill was terribly shaken with one side collapsing—and 'all done in 50 seconds'. …

In the Hutt Valley, the destruction of Alfred Ludlam's house in the Newry. district near the Hutt River was 'momentary'. Ludlam was sitting close to the fireplace with his wife and guests, Charles Bidwill and Mr Hutton. The initial shock, which was associated with a very perceptible vertical movement, seemed to throw the house in the air and shake it. A moment later a second shock sheared the chimneys off their foundations. Ludlam was jammed by a table as he desperately attempted to get up from his chair. Bidwill and Hutton, being closer to the door, grabbed Mrs Ludlam to get her out, but she broke loose and went to her husband just as the main five-metre-high chimney began to collapse. Fortunately the chimney was preceded by a large picture which fell over Ludlam's legs, protecting them both from the bricks and mortar. A fallen lamp gave out its last flicker and they were in darkness. There he and his wife remained trapped. The house continued to be violently shaken—the noise of breaking glass and the sickening feeling of being partly buried in the chimney debris was unbearable. Finally managing to wriggle out from under the mess, Ludlam and his wife reached the hall where they found Bidwill and Hutton holding open the door for them to escape. Now away from the collapsing house, they went to check on the servants and found that they had all climbed out through the windows.

The earthquake was felt throughout the country, with shaking intensity decreasing with distance from the epicentre. In Auckland it was experienced only as 'a slight tremulous motion', but in Hawke's Bay missionary William Colenso clung to the shaking ground outside his house, where he was transfixed by a 'pale lambent fire with ragged blue edges gliding along about a metre from the ground'— it was likely a will-o'-the-wisp caused by 'the spontaneous combustion of methane and other gases liberated by the shaking from the swamp surrounding Colenso's house'. In the Manawatu, two people died when they fell into a fissure. Meanwhile, people in Wellington and other coastal areas were about to experience a tsunami—and the growing realisation that parts of the landscape had changed forever.

About ten minutes after the first great shock, a tsunami swept the coast on both sides of Cook Strait. At Lyall Bay near Wellington, the wave overtopped the Rongotai isthmus and flowed into Evans Bay. The water reached as far as Burnham Water, a lake in the middle of the Miramar Valley, inundating the racecourse and flooding into the gullies beyond, 'to the infinite alarm' of many who were camping there after the races that day. Subsequently, water poured back over the isthmus from the Evans Bay side, carrying with it a lifeboat that was found three weeks later lodged in the scrub-covered sand-dunes near Lyall Bay. The entrance to Wellington Harbour was too narrow to admit the full force of the tsunami, although it was estimated to be about one-and-a-half metres above the highest tide inside the harbour, and nearly four metres higher outside.

In Palliser Bay the tsunami was over nine metres high. Three waves arrived within a few minutes of each other. At Te Kopi, a small but important anchorage for shipping supplies and produce between Wellington and the Wairarapa on the eastern

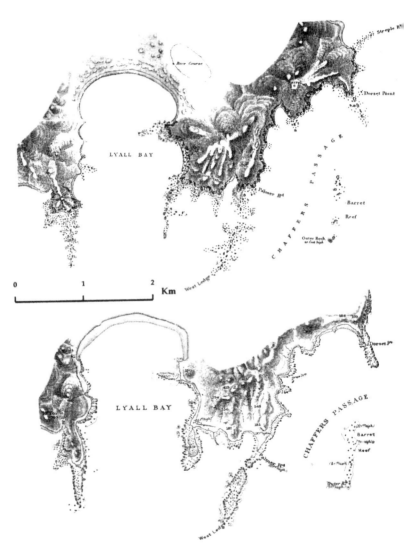

British Admiralty hydrographic surveys of 1849 (top) and 1903 illustrating part of the south Wellington coast and entrance to Wellington Harbour. This comparison clearly shows the effect of the 1.5 to 2 metre uplift of 1855, which exposed coastal rock platforms and caused the seaward extension of the beach at Lyall Bay.

side of the bay, several small whares and their contents, and bales of wool situated on the beach, were washed away. The last of the waves fortuitously dumped the wool bales back on to the beach. Later, the wool was sent on to Wellington just as it was and there were no complaints of damage.

Although the sky was cloudy at the time, the approaching tsunami in Palliser Bay was visible by its white phosphorescent crest. A sailor who had experienced a tsunami on the west coast of South America recognised the nature and danger of the approaching wave as soon as he saw it and was able to warn a boatman's family who lived about eight metres above the beach and would almost certainly have been drowned. The great wave continued north along the east Wairarapa coast causing 'much injury'. On the west coast it flooded into Porirua Harbour and inundated parts of the coastal road between Wellington and Wanganui for many kilometres.

In the South Island the 'gigantic wave' swept the beach near the mouth of the Wairau River and inflicted considerable damage to the 'grog shops' and other shelters built along the Wairau Bar. Along parts of the Kaikoura coast, the sea flooded inland a metre or so above high-water mark. Further south the water level of the Waimakariri River near Christchurch suddenly 'rose 4 feet [1.2 metres] and fell again' and at the Heathcote ferry water washed up the sloping road. A 'wave' came up the Avon River to within two miles of Christchurch and those living along its banks reported hearing the rushing sound of the water as it passed. Next morning the river weeds were found to have been washed on to the grass banks 'for about 1 foot [0.3 metres] in height'. …

As morning dawned, the scene in the town of Wellington was novel. People stood in their doorways, groups sat on mats, or in tents pitched in their gardens, and 'involuntary picnics were to be seen everywhere', despite the clouds of dust from the rubble, which were whipped up by a northwesterly wind. … Those who had been in Wellington during the devastating earthquakes of 1848 remembered that there was another very severe shock 36 hours after the first and were anxiously expecting the same to occur. Many people had pitched tents and others were preparing to spend the night in the bush rather than sleeping in what remained of their houses. … Everyone was comparing experiences and all agreed that the earthquake was of much greater intensity then any they had previously felt. The oldest Maori living in Wellington said they had never experienced anything so severe, nor had they traditions of a like occurrence. They were heard to remark that, prior to this and the earthquakes of 1848, they had 'never thought anything of them'. These previous earthquakes, they said, 'were comparatively slight and never injurious but with each recurrence of these phenomena they appeared to have become more and more severe.' …

One of the first things the inhabitants noticed was the extremely low tide. The beach fronting Lambton Quay extended far beyond its usual limits. This was evidently caused by an uplift of the land by three to four feet (0.9 to 1.2 metres), as attested by the marked difference in the high-water mark before the shock and after it. People gathered in groups to discuss the situation. Many commented on the fact that the ground and gravel on the beach was very hard, as if 'fused'; several others concluded that the upheaval of the land was so great that if they had many more earthquakes like it the harbour would be no more and a strip of land would eventually emerge from Cook Strait, joining the two islands.

The 'beach' along Lambton Quay presented a 'miserable picture'. Few houses had been left undamaged, most were 'perfect wrecks', with the contents of many of the shops floating about in the water or strewn along the shore, as a result of the wave that inundated them during the night. The earthquake had 'disintegrated' parts of the swamp at Te Aro—a morass of flax, raupo and toetoe and small islands of this vegetation were float-ing about in the harbour, presumably dislodged by the wave that had flooded the foreshore immediately following the first great shock. Numerous 'muddy excrescences' had erupted along the foreshore of Lambton Quay. A large totara slab had been placed across the deep fissure that had formed near the corner of Boulcott and Willis Streets and from which mud had flowed down Willis Street.

Earthquake shocks were still continuing but at longer intervals as the day wore on. The earth remained in a state of almost continuous tremulous motion. During some of the more severe aftershocks, people walking in the streets had to stop and hang on to fences. All business was suspended for the day. ...

After surveying the Croisilles area near Nelson, the *Pandora* was back in Wellington on 2 February. Aftershocks were continu-ing at the rate of two or three per day, but were 'slight in degree and seldom longer than 20 to 40 seconds duration'. The towns-people were still hard at work clearing away the fallen debris from the first great shock, but very few had begun to rebuild 'their walls or chimneys'. Jolliffe observed that 'the land in the harbour and all round the coast has been thrown up some feet and in the harbour it is estimated that the rise has been at least three feet [1.2 metres], so that piers and jetties that previously could be reached at the lowest spring tide cannot now be approached within a distance of many yards. The rising of the land is un-doubtedly of benefit to the town of Wellington by giving them considerable additional frontage for building wharves etc.' ...

Prior to the earthquake, the road to the Hutt Valley which ran along the base of steep cliffs was in many places barely above high-tide level. Travellers were frequently in danger from waves during a southerly so the journey was risky. In some places the road had to be retained by a sea wall. Uplift of 1.2 metres caused the road to be elevated beyond the danger of the southerly waves and spared the expense of building further protective walls.

Similarly, before the earthquake much of the Hutt delta was swamp and tidal inlets. The Hutt River was navigable to as far as Maoribank (about 22 kilometres from the river mouth) for smaller craft and for nearly three kilometres, or as high as the Hutt Bridge, for large cargo boats. The Waiwhetu Stream on the east side of the valley was navigable for some distance … The uplift considerably shallowed the Hutt River and Waiwhetu Stream, making them no longer suitable for navigation. Before the earthquake, kahawai could be caught beyond the Hutt Bridge … After the uplift a channel had to be dug for 'more than a mile' [1.6 kilometres] to facilitate the reduced flow of the Second River into the harbour. Wide mud banks were exposed on either side of the Waiwhetu Stream, and a small trading vessel anchored in the stream at the time of the earthquake 'was found high and dry on the mud bank the following morning'. In the vicinity of the Hutt Racecourse it was possible before the earthquake to wade out at low tide to two beached schooners. The uplift left them stranded —the mud-flats where they had been became land dry enough to be laid out as paddocks and ploughed. Much of the swampy area of the Hutt River delta was drained and the shoreline along present-day Petone Esplanade extended out into the harbour. It was common talk among the settlers that the Waiwhetu was used for smuggling goods to the colony on which customs duties could be charged, but with the uplift smuggling abruptly ended.

The Wairarapa earthquake had provided 'the first dramatic proof that faulting … was coupled with uplift of a huge area of land, and that it was directly related to an earthquake not associated with volcanic activity'. Surveys would reveal elevated rock platforms and gravel beaches along the western coastline of Palliser Bay, around Turakirae Head, along the coast to Pencarrow Head at the entrance to Wellington Harbour, all around Wellington Harbour, and westwards along the south Wellington coast. Recent radio-carbon dating across the fault line and at the uplifted beach terraces at Turakirae Head have revealed that the Wairarapa fault has a history of large earthquakes, with an average return period of 1,650 years.

From *Magnitude Eight Plus: New Zealand's Biggest Earthquake* by Rodney Grapes: Victoria University Press, Wellington, 2000.

The Remarkable Pink
and White Terraces

The Austrian geologist Ferdinand von Hochstetter (1829–
84) was one of the first scientists to explore and describe
New Zealand's geology and geomorphology. He arrived
in Auckland in 1858 as geologist to the scientific expedition
of the Austrian frigate *Novara*. The *Novara* set sail again
in January 1859, but Hochstetter remained, persuaded by
Auckland's provincial government to undertake a geological survey. He spent 1859 travelling through the North
Island and Nelson with Julius Haast, a German man who
had also just arrived in Auckland, as his assistant.

Much of the two scientists' time was spent in the
Taupo Volcanic Zone. This account describes the natural
phenomena commonly known as the pink and white
terraces, then touted as the eighth wonder of the world.
In 1886, three decades after Hochstetter's visit, Mount
Tarawera erupted and the terraces were gone, blown apart
by the force of the eruption. This passage is from Charles
Fleming's 1959 translation of Hochstetter's account of
New Zealand's geology. Te Tarata was the boiling spring
that flowed from the top of the white terraces to Lake
Rotomahana beneath.

I do not think that the first impression made by the small dirty
green lake, with its swampy shores and the dreary sombre treeless
hills which surround it, overgrown merely with bracken, is in
any way up to the expectation of a traveller who has heard so
much of the lake's wonders. At least, that is how it was with us.
The lake lacks all the qualities of a beautiful landscape. What
makes it the most remarkable of all New Zealand lakes, one
could even say one of the most remarkable places in the whole

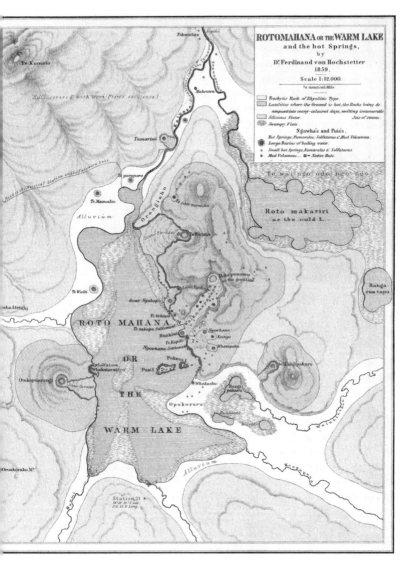

Hochstetter's map of Lake Rotomahana, showing Te Tarata
and the white terraces at the north-eastern end of the lake,
and Otukapuarangi and the pink terraces on the western shore.

MapColl-q830caq/1859–60/Acc.6150–Plate 5, Alexander Turnbull Library,
Wellington

world, must be viewed at close quarters and lies mostly hidden from the newcomer's eyes. Only the steam clouds rising everywhere lead to the suspicion that there is really something to be seen here.

The chief points of interest are concentrated on the eastern shores. There are located the most important of the springs to which the lake owes its reputation, and which are among the most magnificent known in hydrothermal regions anywhere.

At the head of the list stands Te Tarata at the north-eastern end of the lake. This enormous boiling spring, with its sinter terrace projecting far into the lake, is the most marvellous of the wonders of Rotomahana. About 80 feet above the lake, on a fern-clad hillside, from which hot steam escapes in many places that are reddened by iron oxide, the great main basin of the spring lies in a crater-like hollow facing the lake towards the west, surrounded by steep banks of red decomposed clay 30–40 feet high. I estimate it to be 80 feet long and 60 feet wide. It is full to the brim of completely clear transparent water, which appears, in the snow-white sintered basin, a blue of surpassing beauty, turquoise blue, or like the blue of many precious opals. At the edge of the basin I found a water temperature of 84°C (183°F), but in the middle, where the water is in a constant state of ebullition and effervescence to a height of several feet, it reaches boiling point. Enormous clouds of steam, reflecting the beautiful blue of the basin, gyrate and mostly obscure the view of the entire surface of the water, but the noise of ebullition and boiling can always be clearly heard. Akutina (August), the native who served as my guide, said that sometimes the whole water mass is suddenly thrown out with tremendous power; one can then see about 30 feet down into the empty basin, but it fills up again very quickly. Such eruptions are said to occur only during strong persistent east winds. Confirmation of this statement would be of great interest. If it is so, Te Tarata is a geyser that performs at long intervals, the eruptions from which possibly equal in magnificence the famous outbreaks of the Great Geyser of Iceland. The basin of Te Tarata is larger than the basin of the latter geyser, so that the ejected mass of water must be tremendous.

The water is neutral in reaction, has a slightly salty but in no way unpleasant taste, but possesses in high degree the peculiarity of turning objects into stone or more correctly of sintering them over and encrusting them. The deposit is like the siliceous sinter or siliceous tufa at the Icelandic springs, and the outlet of the spring has built up a system of sinter terraces on the slope of the hill, which, white as if hewn from marble, afford a prospect which no description or illustration has power to reproduce. It is as if a waterfall plunging over steps had been suddenly transformed to stone. The idea that a picture can give scarcely reflects the magnificence and peculiarity of the phenomenon in nature. It is necessary to have climbed up the staircase and to have observed the details of structure in order to receive a complete impression of this wondrous edifice.

The extensive low toe of the terrace stretches far into Rotomahana. The terraces begin with minor deposits, which bear shallow pools of water. The further up one goes the higher become the terraces. They are formed of a number of semicircular steps or basins no two of which, however, are of equal height. Each of these steps has a small elevated margin from which a delicate formation of stalactites hangs down on the step below, and a platform, narrow or wide, which contains a basin of water shimmering in the most beautiful blue. These basins of water form natural bath tubs such as the most refined luxury would not have produced more splendidly and comfortably. Basins shallow and deep, great and small, can be selected at will, and at a temperature to one's liking. Since the water flowing from top to bottom over the terraces

Cross section through the basin and sinter terraces of Te Tarata spring: a. main basin; b. basins on the terraces; c. level of Rotomahana; d. siliceous sinter; e. basement composed of decomposed rhyolite.

gradually cools, the lower basins are only lukewarm, whereas those on the higher steps are warmer. One of the basins is so large and deep that one can swim around in it in comfort.

Centuries were needed to form the terraces as we see them today. From the thickness of the deposits of siliceous tufa at the Great Geyser of Iceland, which he assessed as 762 inches, and from the observation that an object placed in the course of the sinter water for 24 hours was covered with a paper thin coating of silica, Forbes has estimated the approximate age of the Great Geyser at 1,036 years. Similar calculations could also be made for Te Tarata, if the thickness of the silica deposits were investigated, and would obviously give a similar great age.

In climbing the steps one is obliged to wade in the lukewarm water, which is spread on the treads of the steps near the deeper basins, but seldom reaches above the ankles. It is not necessary, however, to cope with steaming cascades falling from step to step. Only exceptionally do such occur as the result of more powerful eruptions from the main basin, for generally only a little water trickles over the terraces in separate runnels, and only the principal discharge on the southern side of the terrace structure forms a hot stream with steaming waterfalls. When the highest terrace is reached, the visitor finds himself on a broad platform in which are inset several splendid basins 5 to 6 feet deep, the water of which has a temperature of 30°, 40°, and 50°C. In the centre of this platform, like an island, right on the edge of the main basin, and overgrown with manuka scrub (*Leptospermum*), mosses, lycopods, and ferns, rises a rocky knob about 12 feet high, which can be safely climbed, and gives a view into the blue, boiling and steaming main basin.

Such is the famous spring Te Tarata. The pure white of the sinter formation, the contrasting blue of the water, the green of the surrounding vegetation and the intense red of the naked earth banks of the craters, the gyrating steam clouds—all these combine to form a scene unique of its kind. The collector moreover has ample opportunity to fill whole baskets with fine examples of the most delicate stalactite formations, with encrusted twigs, leaves and the like; everything lying on the terrace becomes in-crusted in a short time. There would be a magnificent field here for 'sinteroplastik', the name given to a young industry originating in Carlsbad, which uses the encrusting properties of calcium carbonate-depositing springs for the preparation of all sorts of pretty objects which have been exposed to the water.

> Hochstetter continues his descriptions of the springs around Lake Rotomahana, finishing with the spring that feeds the pink terraces.

On the west shore, the great terrace fountain Otukapuarangi forms the counterpart of Te Tarata. The steps reach to the lake and the ascent on high is as on an artificial marble staircase, adorned on both sides with manuka, manuwai and tumingi scrub. The terraces are certainly not as magnificent as those of Te Tarata but on the other hand are prettier and finer in their

formation. Moreover, a soft rose red, which invests the wonderful formation as if in a glow, gives the whole an especial beauty. The platform lies about 60 feet above the lake and is 100 yards in length and breadth. It carries pretty basins, 3 to 5 feet deep, full of transparent sky-blue water at 30 to 40°C. In the background, as if in a crater, surrounded by half-naked banks running the gamut of different shades of red, white, and yellow, lies the great spring basin, 40 to 50 feet in diameter and probably very deep. It is a calm, blue pool of water, merely steaming, but not boiling. The water has a temperature of 80°C and the rising steam smells of sulphurous acid. Round about the basin a yellow sulphurous tinge can be noted, and on the side walls of the crater the sulphur has locally been deposited as a thick incrustation. Solfatara activity is shown at its best at the northern foot of the terrace in the solfatara called Te Whaka-Taratara. This is a crater-like basin facing the lake, full of hot yellowish-white muddy water, strongly acid in reaction—a real sulphur lake, from which a hot muddy stream flows into the lake. In the crevices of the banks surrounding the sulphur lake splendid crystals of sulphur can be found.

Photographer Daniel Mundy (1826?–81) took a series of photographs of Lake Rotomahana and its surrounds, which were published in his 1875 book, *Rotomahana and the Boiling Springs of New Zealand*, with descriptive text by Ferdinand von Hochstetter.
Above left: Mundy's photograph of the white terraces, which Hochstetter described as 'a spectacle of superb magnificence [which] form natural baths, such as no human art could have constructed, or fitted in a more luxurious and commodious style.'
Left: The pink terraces, of which Hochstetter said it was 'luxury of luxuries to sit here upon a bed of soft powdery substance, looking round upon the wonderful scenery of Rotomahana, and listening to the cry of the wild birds, on a clear moonlit night.'
BK-822-1 & 2, Alexander Turnbull Library, Wellington

He then talks about two islands on the lake, on which he and his people stayed from April 28–30, 1859.

Puai is a rock pinnacle in the lake not far from its eastern shore, 12 feet high, 250 feet long, and 100 feet wide. Manuka scrub, grass, and fern grow on its surface, and, for temporary visitors to the lake, small raupo huts have been erected in which we settled as comfortably as we could. I think, however, that if we had not known that others before us had already stayed here for weeks, we would, after examining the place more closely, scarcely have decided to spend even a single night on this rocky crag. It is hardly different from living in an active crater. A continuous noise and sizzling, hissing, and bubbling can be heard round about, and the whole soil is warm. In the first night I was terrified because on the soil where I lay in the hut, in spite of a thick foundation of fern and in spite of the woollen blankets which formed my bed, it gradually became so warm from below that I could no longer bear it. I took the temperature. I drove a hole with a stick in the soft clay soil and stuck the thermometer in it. It rose in a flash to boiling temperature. When, however, I withdrew it, hot vapour steamed hissing out, so that I stopped up the hole again in a hurry. In fact, the island is nothing but a porous rock, shattered and riven and decomposed by hot steam and gases, which has been practically cooked soft in the warm lake and threatens to collapse any moment. Hot water emerges around about, partly above, partly below water level. On the south side of the island lies a boiling mud pool, and siliceous sinter blocks strewn about point to great hot springs in former times. In numerous places hot steam still pours out, which we used for cooking, following the natives' instructions. By digging only a little in the ground, or cleaning the rock crevices of the crust that has been formed in them, a ready made oven is produced in which, over a layer of fern fronds, potatoes and meat can be cooked in natural steam. In a few places, the rock fissures are coated with a crust of sulphur, and a strong smell of sulphurous acid makes itself noticeable; elsewhere I found fibrous alum deposited below sheets of siliceous sinter...

In all, about 25 major ngawha can be counted at Rotomahana; I do not even venture to estimate the number of small springs which occur at innumerable points in the area occupying about two square miles. Since these magnificent thermal springs, in the experience of the natives, have proved very curative to chronic skin diseases and rheumatic complaints, it is to be expected that the remarkable lake will become an important watering place and health resort in later years, when the European population has become distributed throughout the North Island.

From *Geology of New Zealand: Contributions to the Geology of the Provinces of Auckland and Nelson* by Ferdinand von Hochstetter, translated from the German and edited by C. A. Fleming: Government Printer, Wellington, 1959.

Julius Haast and Ferdinand von Hochstetter travelled together
in December 1858 and through 1859, undertaking the first
systematic geological survey of the North Island and Nelson.
While they sometimes travelled with a photographer, sketching
was still often the best way to record and describe the landscape
features. In this sketch from Julius Haast's 1858 notebook,
Hochstetter is sitting on top of a hut, sketching the landscape
before him. MS-0921-059, Collection of Sir Julius von Haast, Alexander
Turnbull Library, Wellington

The Great New Zealand Ice Period

Julius Haast (1822–87) arrived in New Zealand in 1858, commissioned by an English shipping firm to assess the prospects for German immigration. He soon joined the Austrian scientist Ferdinand von Hochstetter on a geological exploration of the North Island and the Nelson region. When Hochstetter left New Zealand in late 1859, Haast remained and was engaged by provincial governments to survey the Nelson and Canterbury provinces, which then covered all of the South Island as far south as Mount Aspiring.

A Swiss scientist, Louis Agassiz, had recently promoted the idea that northern Europe's glaciers were the vestiges of an ancient and extensive glaciation, pointing out that landscape features such as striated rock surfaces and morainic deposits could have been created only by glaciers. In his 1879 book *Geology of the Provinces of Canterbury and Westland*, Julius Haast gave evidence to support what he called a Great New Zealand Ice Period.

I have repeatedly alluded to the Great New Zealand Ice Period, or Glacier Epoch, and as this remarkable era plays such an important part in the physical geology of New Zealand, I may be allowed here to make a few remarks upon it for the general reader, the more so as there are scarcely anywhere alpine countries so easily accessible for the scientific observer, bearing such clear and distinct traces of the post-Tertiary Glacier Period, as the alps of New Zealand. The action of the giant ice ploughs, as we may well call these glaciers, has essentially assisted in preparing the lower regions for the use of man, since by it the narrow valleys have been widened, the rugged mountains rounded off,

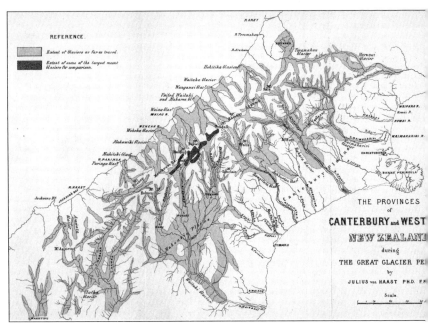

Julius Haast's map of the Canterbury and Westland provinces,
the grey shading indicating his estimation of the extent of
glaciation during the 'Great Glacier Period', or 'Great New
Zealand Ice Period'. The dark shading shows, for comparison,
'the extent of the largest recent glaciers'. B-K 830-370/371,
Alexander Turnbull Library, Wellington

and large plains have been formed. Thus we find everywhere, as soon as we penetrate into the New Zealand alps, where even the outrunning ridges near the plains often attain a height of 6,000 feet, that the valleys are distinguished by rugged forms, where the rivers which break through them have not only cut their bed deeply into the rocks, but have also formed such steep precipices that it is often impossible, even for the pedestrian, to pass along their banks, in order to reach the alpine lakes or plains situated in the valleys above. Inside of the eastern divergent chains, as soon as we enter the district of the earlier post-Tertiary glaciers, the valleys widen out to broad basins, the mountains on both sides— or even standing in the middle of the valley—have the recognised *roche moutonnée,* or ice-worn rounded-hill form, and the fall of the rivers is less rapid. At the upper end of these flats, which are filled up with drift, alluvium, and glacier deposits, and through which the rivers have cut their new bed, lakes, surrounded by distinct moraines, are generally situated. The regular occurrence of these earlier terminal and lateral moraines supply us with the incontestable evidence that these lakes have been formed by the retreat of the glaciers. These lakes are found in every possible stage; some have already disappeared, the delta of the principal tributary entering from the alps having completely filled them up; others are very much contracted by the deltas of the main affluent and of the secondary water-courses descending from both sides; others again are great swamps, having become so shallow, through the enormous quantity of glacier silt deposited in them, that they also may soon disappear under the continually advancing masses of debris.

The extent of these flats, and of the lakes in them, stand in almost all cases in exact proportion to the extent of the present glaciers at the end of the valley, and, therefore, of course to the height, extension, and other orographical conditions of the alpine chains. The form and width of the valleys above the alpine lakes show in the most striking way that they must have once been the bed of great glaciers, to the action of which they principally owe their present form. They are frequently, even up to the present glaciers, of the same width as the lakes. On both sides of them, several thousand feet above the level of the valley, enormous moraines are found stretching along the mountains, so that one can often follow the terminal moraine at the lower end of the lake for twenty miles upwards. *Roche moutonnées* occur everywhere. However, where the colossal glaciers of the Ice Period have pierced through to the Canterbury plains, the secondary ridges are also rounded off and the valleys widened.

In the valleys several miles wide, the present rivers flow in numbers of branches, uniting and separating a hundred times, and changing their bed after each great fresh. At the same time they are often so straight, that one can see the glaciers from the lakes, as is the case, for example, at Lake Pukaki, from which the Tasman glacier, and Lake Takapo [Tekapo], from which the Classen glacier can be seen. Indeed it does not require much power of imagination to bring before one the time when the glaciers were often fifty to sixty miles long and six to ten miles wide. …

In order to understand the former occurrence of the great post-Pliocene glaciers, we have to assume no change in the climate, but simply to consider the existence of the plateau-like mountain systems towering above the snow-line, the enormous snow masses of which might be sufficient to explain the existence and extension of the glaciers. A comparison with the Dovrefield in Norway might perhaps not be out of place here. I have also tried to show that the terrace formation which we meet with everywhere, even in the valley of the most unimportant streamlet, has not been brought about by an upheaval of the land, but only and entirely by the retreating of the sources and the gradual deeper washing out or excavation of the valleys. I have already remarked that most of the principal valleys contract before they enter the Canterbury plains from the lower alps, but some of those which, in the Ice period, lodged particularly large masses of glaciers, form an exception to this rule. I will only quote one as an example: the Rakaia valley, which continues to widen without interruption until it enters the Canterbury plains, where it has attained a breadth of five miles. This is, however, quite natural when we find the last terminal moraine on the Canterbury plains themselves, where it extends over them in a half circle for ten miles, and shows clearly that the ice masses of this glacier were so enormous, that when they came out into the open plains they were able to extend into the form of a gigantic fan. It is, therefore, a matter of course that along the whole valley, from the sources of the glacier to the plains, not only the mountain sides exhibit signs of glacier action and are fringed by moraines of great extent, but that also the mountains in the valley itself must possess the *roche moutonnée* form. Indeed the shape of those in the neighbourhood of Lake Coleridge, a true glacier basin, is so peculiar that they have been named 'sugarloaves,' and the colonists mistook them to be volcanic cones, until I was able to make them acquainted with the real cause of their peculiar form. Thus the Southern Island of New Zealand owes it principally to the Ice Period that, united to the North lsland, it can lay claim to the title of 'Britain of the South', because by its operations have been formed the magnificent plains for agriculture, and the rounded hills and mountain sides so favourable for depasturing cattle and sheep.

From *Geology of the Provinces of Canterbury and Westland, New Zealand: A Report Comprising the Results of Official Explorations* by Julius von Haast: printed at the *Times* office, Christchurch, 1879.

On the Vegetable Food of
the Ancient New Zealanders

William Colenso (1811–99) came to New Zealand in
1834 to run a printing press in Paihia for the Church
Missionary Society. As well as printing thousands of
books and pamphlets, most of them in Maori, he made
many journeys throughout the North Island, on which
he continued the work of the society and indulged his
enthusiasm for botany. Throughout his often chequered
career as a printer, a church deacon, and later a politician,
Colenso continued to collect botanical specimens and
publish scientific articles, including this one from the
Transactions and Proceedings of the New Zealand Institute
in 1880.

The ancient New Zealander had great plenty of good and
wholesome food, both animal and vegetable, but all such with
them was only to be obtained by *labour*, in one shape or the other,
almost unremitting. To them Nature has not been overindulgent
as she had been to their relatives in the more eastern and tropical
isles of the South Pacific—where the breadfruit and the banana,
the coconut and the plantain grew spontaneously, and yielded,
without toil, their delightful fruits to man! But all such constant
labour and industry was doubtless in their favour, helping to 'the
survival of the fittest,' and causing the development of a finer
race, both physically and intellectually. The old Maoris were great
fishers and fowlers—and hunters too, in their diligent snaring of
their prized, fat, frugivorous forest rat; but, for the present, I shall
omit all reference to their animal food, confining myself to their
being industrious and successful agriculturalists and cultivators
of the soil. ...

Of their cultivated food plants, the first in every respect and degree was the kumara. This plant is an annual of tender growth, and was one of their vegetable mainstays. Their use of this plant, as I take it, is from pre-historical times; as their many legends about it evidently show ... In suitable seasons and soils its yield was very plentiful. It had, however, one potent enemy of the insect tribe, in the form of a large larva of one of our largest moths. This larva was named *anuhe*, *awhato*, *hawato*, and *hotete*, and as it rapidly devoured the leaves of the young kumara, it was quite abhorred by the Maoris, who always believed that they were rained down upon their plants. Sometimes their numbers were almost incredible, as some of us have also seen in the abundance of the more common caterpillar pests in certain seasons. I myself have often marvelled at them in their number, and where they could possibly have come from; baskets full being carefully gathered from the plants, and carried off and burnt. This job of gathering them, though necessary, was always greatly disliked. Long before the roots, or tubers, of the kumara were of full size, they were regularly laid under contribution; each plant was visited by old women, with their little sharp-pointed spades or dibbles, who were quite up to their work, and a few of the largest young tubers selected and taken away, and the earth around the plant loosened, when it was again 'hilled' up—an operation not unlike that of our potato hoeing, only much more carefully performed, as at the same time they took away every withered leaf and upper outlying rootlet, and weak sprout. Those young tubers were carefully scraped, and half-dried on clean mattings in the sun—being turned every day and carefully covered from the dew, and when dry either eaten or put away in baskets as a kind of sweetish confection or preserved tuber, greatly esteemed by them, either raw, or soaked and mashed up with a little warm water, and called *kao*.

At the general digging of the crop in the late autumn (called by the Maoris the *hauhakenga*), but always before the first frost, great care was taken in the taking up of the roots, when they were carefully sorted according to size and variety (if of two or more varieties in the one plantation), all bruised, broken, or slightly injured ones being put on one side for early use; then they were gathered up into large flax baskets, always newly made, and in due time stowed away in the proper store; taking great care of doing so only on a perfectly dry sunshiny day, as they had to guard against mouldiness of every kind, which was destructive and dreaded.

It is impossible to estimate, even approximately, the immense quantity of this root which was annually raised by the old Maoris; especially before they took to the cultivation of the introduced potato. At their large and noted tribal feasts, (*hakari*, at the north, *kaihaukai*, at the south) enormous quantities were used, as well as at their commoner feasts held on account of births, betrothals, marriages, deaths etc.; on such great occasions the quantity was often increased through profuse ostentation, for which, while the chief and the tribe gained a great name, they all (especially the women and children) subsequently suffered severely.

But in my opinion one of the most remarkable things pertaining to this useful root, or tuber, has yet to be noticed; namely, its many marked varieties, which were also old and permanent. I have, I think, known more than thirty varieties; and I have lists from the north and the south of several others; and have also heard of others, possibly ten more; while some old sorts were known to have been lost. In this respect the tubers differed just as potatoes do with us. Some were red-skinned, some purple, and others white; some were rough-skinned, and others smooth; some had red flesh, or were pink, or dark purple throughout, others were white; some were even and cylindrical, others were deeply grooved or regularly channelled; some were short and thick with obtuse ends, others were long and tapering with pointed ends; and I never once noticed that there was any mixture (as it were) of the several varieties; all came true to sorts planted, as in the potato with us; their only sign of degeneration through soil or drought was in the size. Now all those several varieties were of old, and only handed down by the strict preserving of the seed (or tuber); and the question with me has ever been, How were they first derived? From the Maoris themselves I never could learn anything satisfactory respecting them—save that they had had them of old from their forefathers. ...

I have carefully enquired if the old Maoris had ever known the kumara to flower, but they all said, 'No; never heard of such a thing.' And they never harvested their crop until after the withering of the leaves of the plant. I have also frequently enquired if any sort or variety had ever been newly raised by them, or their immediate fathers; to this they also replied, 'No.'

Is it not possible that in ancient times this plant did flower here, and that the old cultivators, either by design or accident, obtained their sorts by sowing its seed? The northern tribes, especially the Ngapuhi, had, more than forty years ago, obtained several new varieties of potato by sowing its seed; to which, however, they were first led by accident, having noticed some young plants which had sprung from self-sown seeds of the ripe potato berries, and from them they had obtained several good and prized sorts.

Is it also not possible that this plant (kumara), through constant, assiduous, early, artificial cultivation, extending throughout centuries, has permanently changed in this respect of non-flowering, as it is known the early varieties of potato have done in England through repeated cultivation?

From 'On the Vegetable Food of the Ancient New Zealanders Before Cook's Visit' by William Colenso: *Transactions and Proceedings of the New Zealand Institute*, volume eight, 1880.

Beautiful Plumage

Walter Buller (1838–1906), a New Zealand-born magistrate, lawyer and acclaimed amateur ornithologist, is best known for his book *A History of the Birds of New Zealand*, first published in 1873. This is an excerpt from Buller's ten-page entry on *Heteralocha acutirostris*—the huia—in the book's second edition, which came out in 1888. He includes a portion of a paper he presented to the Wellington Philosophical Society in 1870, in which he described the behaviour of a pair of huia he obtained in 1864. Buller recounts that the birds were caught by 'a native' in the ranges (probably the Ruahine Ranges) and carried on horseback to Manawatu, where Buller exchanged them for a piece of greenstone.

The readiness with which these birds adapted themselves to a condition of captivity was very remarkable. Within a few days after their capture they had become perfectly tame, and did not appear to feel in any degree the restraint of confinement; for, although the window of the apartment in which they were kept was thrown open and replaced by thin wire netting, I never saw them make any attempt to regain their liberty. It is well known, however, that birds of different species differ widely in natural disposition and temper. The captive eagle frets in his sulky pride; the bittern refuses food and dies untamable; the fluttering little humming-bird beats itself to death against the tiny bars of its prison in its futile efforts to escape; and many species that appear to submit readily to their changed condition of life, ultimately pine, sicken, and die. There are other species, again, which cheerfully adapt themselves to their new life, although caged at maturity, and seem to thrive fully as well under confinement as in a state of nature. Parrots, for example, are easily tamed; and I have met with numerous instances of their voluntary return after having regained their liberty. This character of tamability was exemplified to perfection in the huias.

They were fully adult birds, and were caught in the following simple manner. Attracting the birds by an imitation of their cry to the place where he lay concealed, the native, with the aid of a long rod, slipped a running knot over the head of the female and secured her. The male, emboldened by the loss of his mate, suffered himself to be easily caught in the same manner. On receiving these birds I set them free in a well-lined and properly ventilated room, measuring about six feet by eight feet. They appeared to be stiff after their severe jolt on horseback, and after feeding freely on the huhu grub, a pot of which the native had brought with them, they retired to one of the perches I had set up for them, and cuddled together for the night.

In the morning I found them somewhat recruited, feeding with avidity, sipping water from a dish, and flitting about in a very active manner. It was amusing to note their treatment of the huhu. This grub, the larva of a large nocturnal beetle (*Prionoplus reticularis*), which constitutes their principal food, infests all decayed timber, attaining at maturity the size of a man's little finger. Like all grubs of its kind, it is furnished with a hard head and horny mandibles. On offering one of these to the Huia, he would seize it in the middle, and, at once transferring it to his perch and placing one foot firmly upon it, he would tear off the hard parts, and then, throwing the grub upwards to secure it lengthwise in his bill, would swallow it whole. For the first few days these birds were comparatively quiet, remaining stationary on their perch as soon as their hunger was appeased. But they afterwards became more lively and active, indulging in play with each other and seldom remaining more than a few moments in one position. I sent to the woods for a small branched tree, and placed it in the

J.G. Keulemans (1842–1912), a Dutch bird illustrator, drew
this male and female huia for Buller's 1888 edition of *A History
of the Birds of New Zealand*. The male is at the front, with a
short straight beak; the female, behind him, has a distinctive long
curved beak. PUBL-0012-02, Alexander Turnbull Library, Wellington

centre of the room, the floor of which was spread with sand and gravel. It was most interesting to watch these graceful birds hopping from branch to branch, occasionally spreading the tail into a broad fan, displaying themselves in a variety of natural attitudes and then meeting to caress each other with their ivory bills, uttering at the same time a low affectionate twitter. They generally moved along the branches by a succession of light hops, after the manner of the kokako (*Glaucopis wilsoni*); and they often descended to the floor, where their mode of progression was the same. They seemed never to tire of probing and chiselling with their beaks. Having discovered that the canvas lining of the room was pervious, they were incessantly piercing it, and tearing off large strips of paper, till, in the course of a few days, the walls were completely defaced.

But what interested me most of all was the manner in which the birds assisted each other in their search for food, because it appeared to explain the use, in the economy of nature, of the differently formed bills in the two sexes. To divert the birds, I introduced a log of decayed wood infested with the huhu grub. They at once attacked it, carefully probing the softer parts with their bills, and then vigorously assailing them, scooping out the decayed wood till the larva or pupa was visible, when it was carefully drawn from its cell, treated in the way described above, and then swallowed. The very different development of the mandibles in the two sexes enabled them to perform separate offices. The male always attacked the more decayed portions of the wood, chiselling out his prey after the manner of some woodpeckers, while the female probed with her long pliant bill the other cells, where the hardness of the surrounding parts resisted the chisel of her mate. Sometimes I observed the male remove the decayed portion without being able to reach the grub, when the female would at once come to his aid, and accomplish with her long slender bill what he had failed to do. I noticed, however, that the female always appropriated to her own use the morsels thus obtained.

For some days they refused to eat anything but huhu; but by degrees they yielded to a change of food, and at length would eat cooked potato, boiled rice, and raw meat minced up in small pieces. They were kept supplied with a dish of fresh water, but seldom washed themselves, although often repairing to the vessel to drink. Their ordinary call was a soft and clear whistle, at first prolonged, then short and quickly repeated, both birds joining in it. When excited or hungry, they raised their whistling note to a high pitch; at other times it was softly modulated, with variations, or changed into a low chuckling note. Sometimes their cry resembled the whining of young puppies so exactly as almost to defy detection.

> Buller kept the pair of huia for more than a year, awaiting an opportunity to forward them to the Zoological Society of London. Unfortunately, however, through the carelessness of a servant the male bird was accidentally killed, and the female 'manifesting the utmost distress, pined for her mate, and died ten days afterwards'.
>
> While some birds were taken live, it was more common for birds to be shot and traded as skins. Buller obtained most of his bird skins from dealers, but later in his essay he recounts his own October 1883 expedition into a mountain forest in quest of huia. It is astonishing today to read of his delight in shooting huia, but in the late nineteenth century this was accepted ornithological practice. Buller, like many other naturalists, believed that New Zealand's endemic species were doomed, destined to be superseded by 'superior' exotic species. Collecting bird skins was, therefore, considered vital to the scientific description of the species, and to ensure specimens were available for display in museums in New Zealand and Europe.

Taking the early train from Wellington to Masterton on the 9th, I met Captain Mair by appointment, and we forthwith made our arrangements for a start on horseback at daybreak. Instead of a fine day, as we had hoped, the morning opened with a heavy shower, which somewhat delayed our departure, and the day turned out drizzly. Our road lay through a bush and along a highway which had been formed but not metalled. The mire was knee-deep for the horses, and, for most part of the way, it was very toilsome work. The distance to be traversed was only twenty miles, the first four of which were over a hard road; but the shades of evening were closing in around us by the time we reached our camping-ground at the foot of the Patitapu range, and our Maori attendant (Rahui) had barely time to fix up our tent and collect 'whariki' for bedding before thick darkness had set in. Our approach to this camping-place lay along the edge of a wooded ravine. On the opposite side from us there was a grove of tall manuka trees, several hundred acres in extent. Rahui informed us that this was a favourite resort of the huia when feeding on the weta or tree-cricket (*Deinacrida thoracica*). The dull russet-green of the manuka bush was relieved on the sides of the ravine by those ever changing, ever beautiful, light-green tints so characteristic of our New Zealand woods. Here and there a shapely rewarewa reared its tapering top, spangled all over with bunches of crimson flower, while along our path were fringes of the scented pukapuka with its dark green leaves, showing their silver lining as they yielded to the breeze, and covered with a profusion of cream-coloured inflorescence. At intervals might be seen a leafless kowhai laden with a wealth of beautiful golden blossom, and in the more open parts of the widening valley clumps of *Cordyline* with their waving crowns of green; whilst, adding immeasurably to the charm of the whole scene, the star-like clematis, in huge white clusters, hung everywhere in graceful festoons from the tangled vegetation. Down in the bed of the ravine, and hiding the babbling brook, the stunted overhanging trees were for the most part clothed in a luxuriant mantle of kohia, kareao, and other epiphytic plants.

Such was the spot in which we first heard the soft, whistling call of the huia! Rahui imitated the cry, and in a few moments a fine male bird came across the ravine, flying low, taking up his station for a few seconds on a dead tree, and then disappearing, as if by magic, in the undergrowth below. Our guide continued to call, but the huia was shy and would only respond with a low chirping note. But this was enough, and led us to where he was engaged, apparently grubbing among the moss on the ground. We shot the bird, which proved to be in beautiful plumage, and Rahui accepted this as an earnest of our success on the morrow.

Our camp was selected as only a native can select in the bush. The spot fixed upon was a gentle slope under the shadow of a three-stemmed tawhero (*Weinmannia racemosa*), sheltered all round by close-growing porokaiwiria, torotoro, and other shrubby trees, and the whole fenced in, as it were, by a thick undergrowth of bright green pukapuka, mixed with the still brighter mahoe, and protected in front by a perfect network of kareao vines, attached to and suspended from the higher trees. We soon had a roaring camp fire and some ribs of mutton roasting for supper. As the night closed in upon us we heard all round the solemn notes of the New Zealand owl: first, a distinct *kou-kou, kou-kou*; then in a weaker key (perhaps the responsive call of the female) *keo-keo-keo*; and then, in alternation, the alarm-note and the ever familiar cry of 'more-pork'.

Even after a pall of darkness had settled on the woods, some tuis in the tall treetops kept up a delicious liquid song, like the measured tolling of a silver bell, and far into the night could be heard, at intervals, the low whistling note of the kaka communing with his mate. Then all was quiet, the night being very dark, and nothing broke the stillness of the forest till the huia call of our native guide brought us to our senses in the early dawn. But the day turned out unpropitious. The drizzling rain continued and a strong breeze set in; so we determined to shift our camp to the other side of the range. Our road lay along the side of another ravine. We had not proceeded more than a mile when Rahui's call was answered from the other side. The bird's loud cry was presently succeeded by a whistling whimper, and then he came towards us, bouncing through the brushwood as if in a desperate hurry. Descending to the ground a few yards in front of us, he hopped along the surface, and then up the trunk of a prostrate tree, with surprising agility. My companion took a shot at him; but owing to the dampness his gun missed fire, and the bird, taking alarm, disappeared in an instant, all our efforts to recall him proving of no avail. On reaching the head of the valley, we tethered our horses and commenced the ascent of the range, which we found very steep. About halfway up, we rested on the ground. Rahui continued his call—a loud clear whistle—not much like the ordinary call of the bird, being louder and more shrill. In a few seconds, without sound or warning of any kind, a huia came bounding along, almost tumbling, through the close foliage of the pukapuka, and presented himself to view at such close range that it was impossible to fire. This gave me an opportunity of watching this beautiful bird and marking his noble bearing, if I may so express it, before I shot him. While waiting to get the bird within proper range, I heard far below me the rich note of the kokako, repeated several times. It is scarcely distinguishable from the call of the tui, but is preceded by a prolonged organ note of rare sweetness. My next shot was at an adult male huia who came dashing up, with reckless impetuosity, from the wooded gully.

Being anxious to obtain a perfect specimen, I risked a long shot and only wounded my bird. Down he went to the ground like an arrow, with a sharp flute-note of surprise or pain, and then darted off, kangaroo-fashion, covering the ground with wonderful rapidity, and disappeared in the tangle.

We found the descent of the range much easier than our toilsome climb. Remounting our horses we continued up the valley. At a turn in the road, at a spot hemmed in by a wooded amphitheatre of beautiful shapely trees (chiefly rata), we halted for a moment to gaze on the scene. On a tree, immediately in front of us, a pair of wood pigeons were sitting side by side, show-ing off their ample white breasts under the rays of sunlight glancing through the rain-drops. Whilst we were looking at and admiring this little picture of bird life, a pair of huias, without uttering a sound, appeared in a tree overhead, and as they were caressing each other with their beautiful bills, a charge of No. 6 brought both to the ground together. The incident was rather touching, and I felt almost glad that the shot was not mine, although by no means loth to appropriate the two fine specimens. Before we reached our next camping ground, at the foot of Poroporo, we had bagged another bird (a female of last year) who was unattended, and came up quite fearlessly to her doom.

After we had secured our horses and 'refreshed the inner man', Rahui and I started again for huias, whilst our companion re-mained to fish for eels in the creek near our camp. After we had walked about a mile, a bird answered our call, and immediately afterwards a pair of huias alighted in a pukatea tree above us. I brought them down, with right and left, and then another bird (a young male) appeared on the scene. He exhibited great ex-citement and was evidently at a loss to know what it all meant. Uttering a low, sibilant cry, with a tender pathos, he hopped down lower and lower, till within a yard or two of my head. I could easily have knocked the pretty creature over with a stick, but had not the heart to do so. I was less scrupulous, however, about having him caught, and in far less time than I take to write it, Rahui had selected a long stick, fixed a noose at the end of it, and slipped it over the bird's head. The huia nimbly jumped through the loop but was caught by the feet. On finding himself a captive, he uttered no sound, but, in the most practical way, at once attacked my hands with his bill, striking fiercely and repeatedly at a white-faced signet-ring. On the following day Rahui managed to snare another, which was fortunately a female, thus making a pair of young birds. They became at once reconciled to confinement, eating freely of the huhu grub, and resting very contentedly on a perch to which they had been attached by a thong of flax. The young of the first year has a low and rather plaintive cry, easily distinguished from all other sounds in the forest, and pleasant enough to the ear. Our third and last day turned out wet and stormy; but we nevertheless got some more huias, our bag consisting altogether of sixteen birds, exclusive of the live ones.

In 1891 Buller had a change of heart, and supported plans for statutory protection of the huia and for the creation of sanctuaries on Resolution and Little Barrier Islands. However, despite the 1892 amendment of the Wild Birds Protection Act to include the huia, the bird would continue to fall victim to extensive habitat destruction, introduced predators, and collectors' greed for skins, not just for scientific study but for display in fashionable drawing rooms and as a source of decorative feathers. The last sighting of the huia was in 1907 and the bird Buller described as 'graceful' and 'shy' with 'beautiful plumage' is now extinct.

From *A History of the Birds of New Zealand* by Sir Walter Lawry Buller, second edition, volume one, published by the author for the subscribers, London, 1888.

Huia

The huia, once widespread in the North Island, became extinct in the early twentieth century, following decades of habitat destruction, predation by rats, stoats, ferrets and cats, and hunting by humans.

Huia feathers were always rare treasures,
kept in waka huia,
treasure boxes.

An iridescent bird, blue-black like petrol,
with a greenish sheen,
rarely seen,

the huia hopped along the ground, grounded.
But sang like the tui,
like a flute.

Dressed in treasures too valued by people
for the bird to be valued as bird,
the huia is no longer heard.

When the Duke of York was presented
with a feather for his hat,
trade

in huia feathers leapt to extinction.
Now the waka huia preserve
other treasures.

This is my waka huia for the bird.

From *The Pastoral Kitchen* by Anna Jackson: Auckland University Press, Auckland, 2001.

Search for the Stitchbird

Andreas Reischek (1845–1902), an Austrian naturalist, came to New Zealand in 1877 as taxidermist to the Canterbury Museum. After fulfilling his duties at the museum, Reischek travelled New Zealand with his dog Caesar, working on his own collection of natural history specimens, much of which eventually made its way to the Museum of Natural History in Vienna. In the 1880s he made several trips in search of the stitchbird, which he called the ti-ora. Reischek wrote of his Antipodean adventures in *Yesterdays in Maoriland*, the English translation of which was first published in 1930.

You may be surprised that any man, instead of … devoting himself to wife and family, should go running after a rare bird that nobody had seen for years, and give up all his hardly gotten gains to such a purpose. But from the day I saw the first stuffed specimen of the *Pogonornis cincta* in Christchurch Museum, and learnt from Sir Julius von Haast that a few examples were still said to live in the virgin bush of one of the most mountainous islands of the Hauraki Gulf, I resolved to seek him out, or die in the attempt.

I was to find I needed all the energy and perseverance I could muster before I was able to locate this practically unknown creature, this bird of wonder, the ti-ora of the Maori. At last, after months and months of patient search, after traversing every part of this rugged island, and climbing up and down ranges 2,000 feet above the level of the sea, in the deep and silent recesses of the Hauturu bush it suddenly appeared before me—like the blue flower of romance which at length crowns the efforts of the believing seeker.

I started for Little Barrier—as the colonial calls Hauturu—in October 1880, accompanied by my friend, Mr E. Firth. We left Auckland by the schooner *Rangatira*, belonging to Chief Tenatahi, who piloted it himself. We struck a severe storm off Kawau, which tossed our little vessel up and down like a nutshell, and when towards evening we got near Hauturu, it was to find landing impossible on account of the breakers. We therefore put into Little Omaha to wait for better weather, and remained there two days, the natives amusing themselves diving for lobsters, which are very numerous here. Many a hand and face was scratched in the process, but that did not prevent further diving.

The third day we steered for the north-west of the island, where we effected a landing after one or two vain attempts. Bedding and stores were unloaded at a deserted native hut not far from the shore, and then the cutter sailed away before night came on. I spread my blankets out on the floor and tried to sleep, but a European goblin, which had established itself in numbers, gave me no rest. The camp was swarming with fleas! There was nothing for it but to turn out into the open till next morning, when I gave the place a thorough clean out.

Then my usual trail-blazing began. The island was precipitous and overgrown with luxurious bush, consisting principally of giant kauri trees (*Dammara*) reaching skywards like mighty cathedral towers, while below the dark green arches of manuka and nikau palms, and the tender soft green veil of broad fern-tree fronds, were richly contrasted.

I observed many kinds of birds, but the ti-ora was not one of them. Alas! it was neither to be seen nor heard: and after several weeks of fruitless search I gave it up for the time being. It was through the western and south-western parts of the island that I searched. I intended penetrating into the centre, but the natives told me that it was impossible to get overland to the east coast on account of the many precipices, and that the sea was too rough to permit of my landing on that side; so I returned to Auckland with the intention of resuming the search at another time.

In May 1882, I sent Dobson, my assistant, to Hauturu for the purpose of repairing old huts and building new ones, taking provisions for a prolonged mountain expedition, my intention being to follow him in June. I did indeed leave Auckland that month in the cutter *Water Lily*, but the weather proved so bois-terous that after two attempts to land, and five days battering on the open sea, we had to give it up.

The following month I made another attempt on the *Ranga-tira*, but with no better success, though on this occasion we only stuck it out for three days; so I put off my trip until October, on the 15th of which month I really did succeed in landing. Dobson was waiting for me on the rocks, and after packing all my kit into a little boat, we towed her round to the south-east side. Here we carried our things ashore and dragged the boat up, as we thought, well out of the sea's reach.

That night we camped at the foot of a precipice, ate our primi-tive supper of ship's biscuits and water, and lay down to rest. A glorious night arched over our solitude. A grey mist crept silently over the sea's face, and above our heads millions of stars came out, twinkling like diamonds in a dark-blue sky. The roar of the breakers surrounded us like a mighty organ, playing the lullaby of an unspoiled world.

In this primeval paradise I felt the windows of my soul were opened, Nature's wonderful mantle lay spread out before me; as never before I realised the spiritual kinship between all living things, the connection and coherence of the manifold works of God. That night, lying there, I experienced a sense of shame, which those who swear by civilisation will certainly fail to under-stand, that civilised man can be the worst vermin of the whole

Earth. For wherever he comes, he destroys the wonderful equipoise of Nature, and much as he bothers himself with his so-called arts, he is not even capable of repairing the damage he causes. As if in sympathy with my thoughts, about three o'clock a comet appeared in the sky, its long tail glowing with a pallid light.

Before dawn we commenced the formidable ascent. To give some idea of the difficulty of climbing here, I had to pull my dog—an experienced alpine climber—up after me with a rope, then our gear and provisions, and finally Dobson. From the top we struck off valleywards, and up and down again, over two ranges, each above 2,000 feet high, till towards evening we arrived at an old nikau whare, which my friend had previously built, at the foot of the last range.

It was dark before we finished mending the roof and preparing for a start the next day. On the morning of the 23rd, I first heard the whistle of the ti-ora. I was unable, however, to get a glimpse of it; and though we cut tracks to the tops of most of the main ranges, and afterwards frequently heard the birds, we could never see them. Later experience taught me that their shrill whistle is very deceptive, and the sound travels a long distance.

I then shifted my quarters farther towards the interior; and on the 25th, while again wielding a slasher, my attention was arrested by the call of my dog at a short distance. On going towards him, I saw a male ti-ora hopping about in a very excited manner in the scrub above him. I was thrilled to the marrow by my first view of this magnificently coloured rarity, which has a brighter plumage than any of its New Zealand compeers, and stood watching its quiet and graceful movements without attempting to use my gun.

After this, though constantly exploring, I never saw another specimen till November 7, yet frequently heard them. Early that morning we travelled north-west to the top of a high, narrow

Illustration of ti-ora, the stitchbird, from Andreas Reischek's *Yesterdays in Maoriland*. Reischek searched for this bird for many years. When he found a population he shot 150 specimens.

range of precipices, overgrown with short, thick scrub and manga-manga, which made it so dense that I had to cut the way with my hunting-knife.

This place I found a favourite resort of these birds—which had cost me so much time, labour, and patience—for it had a warm aspect, exposed to the sun. There I saw a male and a female, the latter for the first time; and on the following day I saw a male at the same place. On going over a range, I heard another. Subsequently I went round it, and saw male and female near a nest, and endeavoured to observe them unnoticed, but they quickly saw me, and in the act of escaping I shot them. I then went and examined the nest, which was only half finished, built of very small branches, roots, and fine native grass, and lined with hairy substance off the fronds of the punga.

When our provisions were nearly exhausted, we made our way down to the ocean, only to find our boat was missing. After a long search we spotted bits of broken planks and the rudder among the rocks. In spite of the high position we had dragged it to, the waves had reached it during a storm, and battered it to pieces.

There was nothing for it now but to clamber along the wild coast, humping our packs. In this way we managed after dark to reach the native settlement, where we stayed for a week, and then returned to Auckland by the *Rangatira*.

In 1883, summer being well under way, I again went to Hauturu, and had even more success in observing the ti-ora. In December of that year, I once more took possession of my little hut in the centre of the island, and setting out one rainy morn, I observed a pair of adults with three young birds. On the male noticing me, he uttered a shrill whistle, and the female immediately hid amongst the fern for a considerable time. I went partly at the request of Sir Walter Buller, for whom I procured specimens of which his collection was deficient.

To my great joy I found this rare bird had increased since my last visit, which I put down to the fact that I had on that occasion shot a number of wild cats and the older male birds. I was able to watch whole families of them, and discovered that the young birds possess an intermediate plumage, especially noticeable in the males. I have only once seen these birds sitting still, and that was near the nest. They appear always on the move, carrying their heads proudly, their wings drooped, and their tails spread and raised; and at each successive movement they utter that peculiar whistle from which the natives have named them ti-ora. The female has a different note, sounding like 'Tac, tac, tac!' repeated several times.

They feed on small berries and insects, and suck the honey from the native wild flowers and trees, as many of the latter exude honey during the night. They are not strong on the wing, but very active in hopping and climbing, which enables them quickly to escape from sight.

The plumage of the male is as follows: head and neck, shining velvet black, with a few long silvery-white ear feathers; shoulders, golden yellow; upper secondary, white, with brownish-black points, and a slight splash of white under the wing covers; wings and tail, brownish black, each feather edged on the outer side with olive green; tail cover, greenish tinge, and a yellow band round the breast; abdomen, grayish brown; bill, black; eyes, dark brown; feet, light brown. The female is a little smaller, of olive brown colour on the top of the head, back, wing, and tail; ear feathers hardly perceptible, and a few other differences I won't mention. So far as I know, the plumage of the young, which differs from the adult bird, has never been described.

On December 16 I climbed to the topmost peak of the island, where a heavy thunderstorm surprised me. Flash followed flash, and the thunder rolled formidably below me through the ravines, which soon became raging torrents, through which I had to struggle back to my nikau whare through a regular hail of stones and falling branches.

Close to the hut I heard two miro call, and going outside I found them anxiously hopping about on a manuka clump. I approached, and discovered their water-logged nest containing three eggs.

On the 19th I got my collection together, and made for the Maori settlement, where Chief Tenatahi related to me how he had lost his cutter *Rangatira*, with which he had previously won several races at Auckland Regatta. He had sailed to Catherine Bay to extract blubber from a whale which the natives had caught, and had been overtaken by a storm while at anchor there. In trying to run out, the anchor got wedged, and he and the crew had to jump overboard to save their lives, while their boat was battered to pieces on the rocks.

Tenatahi, his wife Rahni, two men, and a boy then set off in a whaling boat for Tiharea, their Hauturu settlement, but some miles off the Great Barrier another storm came on. The boat turned turtle, and they lost an oar, whereupon Tenatahi ordered the men to keep her steady while he righted her, and baled her out. Rahni, who was a good swimmer, jumped in and swam after the oar, but by the time she brought it back, one of the men had been washed away. They had a frightful job to get the second man and the boy back into the boat, and the latter died soon after from exposure and cold. The man, too, would have died if they had not made him row for all he was worth to get warm.

Three times after this the boat filled with water, and but for Rahni's skill and courage all would have been up. She was fourteen hours in the water without food or drink, battling with the waves, and when they finally reached the shore, was so exhausted that she could not move a foot. Rahni, I may say, is 5 feet 10 inches tall, and the possessor of a fine if muscular figure.

I paid my last visit to Hauturu on April 8, 1885, to procure specimens for the use of New Zealand museums. I found my old hut had been burnt down by Maori youngsters, so I went to live at their flea-infested settlement. In spite of not feeling too well, I pursued my observations unceasingly.

Again in the centre of the island I was successful in observing a pair feed their young (two males and a female) which must have been a late brood. I fear these very rare birds will soon disappear, even from these lonely depths of the bush where human beings had never hitherto penetrated, largely owing to wild cats, which have become very numerous and commit great havoc among them, and also the sparrow-hawk (*Hieracidea novaezealandiae*) and the little morepork owl, in whose crops I have often found their remains.

I returned to Auckland in the middle of May; but before I left, Tenatahi, the owner of the island, invited me to a ball in the native runanga. All the inhabitants of the island were present, mostly

Maoris, with two Portuguese and two white girls to represent the white race. Dance music was provided by a Maori playing waltz, polka, mazurka, and quadrille music on an accordion. The polyglot chief sang snatches of English, Maori, Portuguese, and even German songs, and about midnight we sat down to a supper of bread, honey and tea.

> In December 1880 Reischek visited the Hen and Chickens Islands in Northland where—with the help of Caesar—he discovered the unusual living habits of the *Puffinus assimilis*, or little shearwater, a type of petrel.

A remarkable fact was that I found this bird living in burrows in common with the tuatara, the rare and singular fringe-back lizard, and the last remaining offspring of the Saurians. I usually found him in the first chamber from the entrance, and the bird in the second. These two dissimilar companions live together in harmony, and mutually protect their young, as I had reason to remember, suffering bodily as a result of their knightly friendship.

Caesar announced the discovery of one of their holes with a bark. Plunging in my hand, I felt the bird inside, but was unable to catch hold of it. I therefore took off my coat to enable me to reach farther in, and was groping about when something bit on to my thumb and hung on to it. With my free hand I got out my sheath-knife, and widened the hole, and when I was able to draw out my trapped hand, it was to see a big lizard hanging to my thumb. He only let go when I caught him by the neck.

Inside I found, to the left, the petrel sitting on one egg in a hollow strewn with leaves and grass, and to the right, the lizard's nest. I discovered several more such holes, and everywhere was the same arrangement.

The lizards pass the whole day in their holes, and only come out in the evening to hunt for insects. While doing this, they utter

Reischek found tuatara living in the same burrows as the little shearwater. This illustration appeared in his book *Yesterdays in Maoriland.*

croaking noises similar to those of a frog. As the colour of their skin matches that of their surroundings, they are only to be found with difficulty. On discovery, they do not slip away like other lizards, but stay quite still. Only if one happens to step close to them, they quickly dart into a hole, and if one tries to hold them, they defend themselves with bites and scratches.

On all the larger islands they live principally on insectivorous foods, such as beetles, grubs, wetas, grasshoppers, flies etc. which I found on dissecting. I think where such food—which they prefer even in confinement—is plentiful, they will not prey on birds.[*]

---

[*] More recent studies suggest tuatara eat petrel eggs and chicks, as well as flies attracted by the remains of the petrels' meals.

But on my visit to Karewa Island, at the beginning of 1885, I saw many young birds with their heads off, and I followed one of these lizards with a bird of considerable size in its mouth, which tried to escape in a burrow, but got stuck at the entrance. They catch the bird by its head, and then chew until it is devoured. My opinion is that, as this island is small, and these lizards so numerous, this is the reason they prey upon birds.

As for the puffins, in the daytime only single specimens and their young remain on the island, but in the evening we saw flocks of thousands circling round the camp. They seemed rather surprised to find a solitary habitation occupied. After sunset they settled on the ground, in some places so thickly that one could hardly walk without treading on them; instead of going out of the way, they defended themselves by biting. They even came into our tent, and we were obliged to throw them out and shut it up; then they burrowed in underneath.

When preparing tea, one of us had to watch and keep them off the fire, and, when frying fish, they actually walked into the frying-pan. The variety of their vocal powers was most amazing, and when they joined in chorus it was deafening. One night I went into the bush with a light, for the purpose of observation; a whole flock of these birds flew at me, and knocked the light out of my hand. I did not allow my dog to touch them, and they went on his back, walked over him, and sat alongside him. They are, however, very vicious when molested.

From *Yesterdays in Maoriland: New Zealand in the 'Eighties* by Andreas Reischek, translated and edited by H.E.L. Priday: Jonathan Cape, London, 1930.

The Skeleton of the Great Moa in the Canterbury Museum, Christchurch

The British ornithologist Richard Owen was, in 1840, the first scientist to identify the *Dinornis novaezealandiae*, a species of moa. In the next decade, collectors gathered and packaged up many crates of moa bones to send to Owen at the British Museum, where he identified what he believed were thirteen different species of moa. Once museums were established in New Zealand, they too began to collect moa bones and reconstruct skeletons for display. This poem by Allen Curnow was first published in 1943.

The skeleton of the moa on iron crutches
Broods over no great waste; a private swamp
Was where this tree grew feathers once, that hatches
Its dusty clutch, and guards them from the damp.

Interesting failure to adapt on islands,
Taller but not more fallen than I, who come
Bone to his bone, peculiarly New Zealand's.
The eyes of children flicker round this tomb

Under the skylights, wonder at the huge egg
Found in a thousand pieces, pieced together
But with less patience than the bones that dug
In time deep shelter against ocean weather:

Not I, some child, born in a marvellous year,
Will learn the trick of standing upright here.

From *Early Days Yet: New and Collected Poems 1941–1997* by Allen Curnow: Auckland University Press, Auckland, 1997.

After the Tarawera Eruption

Mount Tarawera, a volcano 24 kilometres south-east of Rotorua, erupted violently and unexpectedly on June 10, 1886, killing more than one hundred people, destroying the world-famous pink and white terraces of Lake Rotomahana, and forever changing the surrounding landscape. Assistant surveyor-general Percy Smith (1840–1922) and analytical chemist James Pond (1846–1941) reported on the eruption —signs of which they had witnessed from Auckland—after visiting the area from June 13 to 16.

The 10th of June, 1886, is likely ever to be remembered in the history of New Zealand as that on which the colonists first had practically brought home to them the fact that the volcanic forces for which these islands are so celebrated had still an amount of vitality in them that was unlooked for and unexpected. The eruption of Tarawera Mountain, and the conversion of Rotomahana Lake into a crater, on that date, at about 2.15 a.m. has caused widespread consternation, the loss of several lives, and a feeling of anxiety as to whether this outburst will be confined to the immediate district where it occurred, or whether it will spread to others in which the signs of thermal action have been known for long periods. …

The amount of information which has been recorded as to the actual outburst is very considerable, but … the best accounts obtainable seem to place the first signs of anything extraordinary happening, at about 1 a.m. on the 10th June, 1886, when slight earthquake shocks were felt by the people at Wairoa, and at Rotorua, (accompanied at the latter place by rumbling noises), which appear to have been continued as earth-tremors till 2 a.m., or past. At 2.10 or 2.20 the rumbling noise had become a

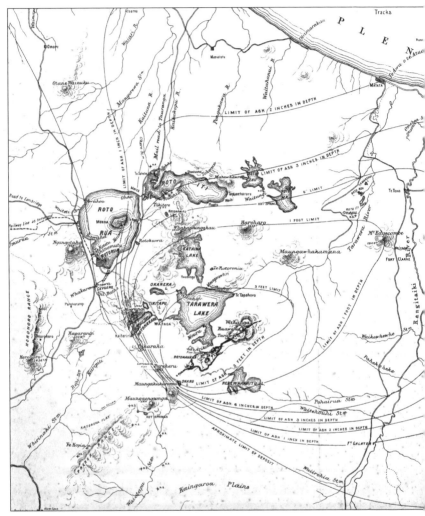

A detail of the map of the district affected by the Tarawera eruption, showing the extent of ash deposits from 2 feet deep (0.6 metres) around the eruption zone, to 1 inch (2.5 centimetres) deep at the coast near Te Puke. This map was drawn by A. Harding to accompany a report by A.P.W. Thomas, professor of natural science at Auckland University College, who, like Smith and Pond, reported on the eruption and its effects. MapColl 832.18cbh/1887/Acc.50472, Alexander Turnbull Library, Wellington continuous and fearful roar, accompanied by a heavy shock of earthquake; and at this same time, or immediately afterwards, an enormous cloud of smoke and vapour was observed from Wairoa, rising over the hills which shut in that village from a clear view towards Tarawera Mountain, the outside edges and fringes of the different masses of which were outlined by vivid flashes of electricity, darting through the cloud and colouring it most brilliantly and beautifully. This electric display was accompanied by a rustling or crackling noise, which appears to have been heard above the deafening roar, and which is probably the same noise as is heard in electric discharges of an artificial kind, and also probably the same as is heard sometimes at great auroral displays. … It was noted by two observers, (Messrs Blythe and Greenlees), that from 2.30 onwards severe shocks occurred at regular ten-minute intervals up to 3.30. The latter gentleman had the presence of mind to observe, from the swinging of a ham, that the shocks came from the direction of Tarawera. … As described by Mr McRae, who saw it from the old Mission Station, soon after the outburst, three columns of fire and flame (or probably the glare reflected on the vapour from lava below) were shooting upward from the flat plateau-like summit of the mountain to an immense height, with flashes of electricity darting forth in all directions, accompanied by balls of fire, some of which fell at great distances, indeed as far off as the Wairoa village, some eight miles from the seat of eruption. Small stones now began to fall, as the great black cloud which had formed over the mountain worked towards the west, to be quickly followed by a downpour of mud and water and heavy stones, which battered down many of the houses in the village. The mud appears to have fallen in the form of an exceedingly heavy rain, with sometimes large lumps of mud, and this continued up till 6 a.m. All this time, there appears to have been a more or less strong odour of sulphur experienced by the people at Wairoa; and

Mr Blythe describes a hot suffocating blast, which nearly choked himself and Miss Haszard, after their escape from the burning house, and which warmed them through.

Soon after the first outburst, and before the fall of the first stones, a great wind arose, which rushed in the direction of the point of eruption with great force, and was most bitterly cold. It is noticeable that the people who survived, and were nearest to the seat of the eruption, viz. those at the Wairoa, failed to hear the loud detonations which reached Auckland and other places. Probably the loud and continuous roar drowned the louder reports.

These explosions ... sounded like the reports of distant cannon, or—as has been described by a large number of people from different places—like someone banging an iron tank. The flashes of the electric display were distinctly seen here in Auckland, a distance of 120 miles in a straight line from Tarawera. The immense cloud of ashes, mud, and sand which was shot high up into the air darkened the sky till long after daylight should have appeared. ... The height to which the mass of light ashes was ejected must have been enormous. ... We know from actual measurement that the column of steam arising from Roto-mahana several days after the eruption was 15,400 feet, and even then the top of the column could not be seen, from its proximity to the observer. The ashes and dust ejected fell on the coastline at points 160 miles apart in a straight line—viz. at Tairua and at Anaura, a few miles north of Gisborne, and some of it fell on the SS *Southern Cross* off the East Cape, and on the SS *Wellington* near Mayor Island. It thus covered an area of land equal to 5,700 square miles with more or less of the deposit; on the edges of which, of course, it is barely visible. ...

The electric phenomena accompanying the outburst must have been on the grandest scale. The vast cloud appears to have been highly charged with lightning, which was flashing and darting across and through it: sometimes shooting upwards in long curved streamers, at others following horizontal or downward directions, the flashes frequently ending in balls of fire, which as often burst into thousands of rocket-like stars. Fire-balls fell at the Wairoa and other places, and doubtless the fires which occurred at Mr Haszard's house and in the forest near Lake Tarawera were due to these. …

The hot springs in the neighbourhood of Rotorua were greatly affected. A small steam fumarole … near the Government Agent's house became a large boiling spring about ten feet in diameter, from which a good-sized stream of hot water ran away towards the lake. Further north—at the base of the Pukeroa hill, and in the direction of the Maori village of Ohinemutu—steam came forth from innumerable cracks in the earth, sometimes accompanied by hot water, which formed streams running alongside the road from the old to new township; and in the pa itself a spring burst out in the great meeting house of Tamate Kapua; another in the path leading down to it; and yet another just behind the building. All of these outbursts occurred on the night of the eruption; they all follow, however, the old deposits of sinter at the base of the Pukeroa hill—the last remaining signs of former great activity in that locality. The activity of the vast number of fumaroles and springs in and around Ohinemutu was certainly greater than usual a few days after the 10th. …

Smith and Pond arrived in the eruption zone just three days after the event.

Emerging from the bush … the scene is wonderfully striking. The whole country is clothed in a pale grey mantle. Hill and dale, level and steep, all is of the same hue. In the far distance, as in the near

Percy Smith and James Pond's June 1886 visit to the eruption zone was followed by a more formal topographic survey by Smith and other survey department staff in July. Photographer Charles Spencer accompanied the team for a few days, recording their work in the drastically changed landscape. *Top:* Smith and his team among the ruins of McRae's Hotel, Wairoa. *Bottom:* Smith and his survey party approach the north-west edge of the Rotomahana crater. F-2516-½ & PA1-0-122-63, Crompton-Smith Collection, Alexander Turnbull Library, Wellington foreground, nothing has escaped this ashen covering save the Okaro Lake, which lies before us sombre, silent, and unruffled. Away in the front rises an ever-rolling, slow-changing, towering mass of steam, interspersed in the lower portions with sudden bursts of darker material, which prove to be stones, sand, mud and water, flung up to the height of 400 or 500 feet above the lip of the crater. At times, the bright sun glancing over this wondrous column gives a vivid brightness to it; and again, so brilliantly reflected is the sunlight from the more distant portions of the mantled earth, as to bring vividly to the mind of the onlooker the semblance of a vast field of snow.

On entering this sombre plain, the ashen covering proves to be a fine, dry, powdered material, having throughout small fragments of scoria. Occasionally spherical or ovoid nodules are found, which easily crush between the fingers, and sometimes contain a nucleus in the shape of a rounded fragment of scoria.

Advancing through this material—which closely resembles in colour and appearance Portland cement—the deposit becomes deeper, so that walking was very fatiguing. In many parts each step was knee-deep, while, by leaving the ridges, the soft ash was found to be so deep as to be dangerous, and the effects of the wind stirring the surface made breathing laboured.

Travelling somewhat to the north of Okaro Lake for the distance of about a mile and a half brought us to the most southern part of the fissure, which has extended from the Rotomakariri Lake in the direction of the Okaro Lake, partly through the Haumi Stream. On the line of the fissure in this direction are five distinct craters, the most northerly of which was decidedly the most active, while the southerly one was nearly dormant. ...

One looks in vain for any sign of the Pink Terrace: all view in that direction is cut off by the column of steam. The edge of Roto-mahana Lake is now far within the crater wall, which follows round from our immediate front in a westerly, then north-westerly, northerly, and north-easterly direction to the site of the White Terrace. The crater has clearly eaten its way back from the edge of the lake, a distance of at least a quarter of a mile from the site of the Pink Terrace; and all along the foot of the wall the steam rises from so many points that it is impossible for the eye to penetrate within its precincts, except on rare occasions when the wind causes a separation of the masses of vapour; and then is disclosed to view for a short time a cavernous-looking aperture, in which can be discerned a picture once seen never to be forgotten. A dismal coffee-coloured light, penetrating the vast mass of vapour from above, enables us indistinctly to see a horrible mass of seething, boiling waters, stained of a black or dirty brown colour, encircled by walls and hillocks of dreadful-looking hot mud, from which the steam curls up in innumerable places. Mud volcanoes scatter their contents around on all sides, whilst every now and then a loud detonation precedes the discharge of a column of water, mud, and stones high into the air, and as they fall splash the black mud right and left. The whole interior surface of the crater, as far as the eye can penetrate, seems to have been boiled and steamed and hurled about to such an extent that the old landmarks are no longer recognisable. Whilst the greatest activity seems to follow the foot of the crater wall round by the western side, the eastern has also its points of eruption, from which vast columns of steam arise to join the general mass above; but, as yet, no one has been able to obtain a clear view of this eastern side. The size of this crateral hollow is about one and a half miles in a north and south direction, with a width of about one and a quarter miles.

From a point which was reached with great difficulty on the west side of the crater, a view is obtained looking north-east, past the site of the White Terraces, and embracing the whole of Tarawera Mountain. The deep sand in this direction makes progression most slow and fatiguing, and not without danger from the slips of sand on the steep hill sides. We looked in vain for any sign of the White Terraces; and as the eye gradually got to recognise some of the more prominent features of the country near there, under their altered shapes and appearance, the conclusion was forced on us that these beautiful terraces—the most lovely and wonderful of their kind on the whole Earth—had disappeared for ever from mortal view. The changes in the general appearance of the country near there are so great, that, even with a familiar knowledge of the locality, which had been impressed on the mind in a visit to the same spot on which we now stood only three short months before, we recognised with great difficulty and uncertainty the main features of the land. But, still, the evidence of the whole contour of the country goes to show that the site of the terraces is now occupied by a horseshoe-shaped recess or bay in the general line of the main crater, from which an enormous column of steam arises high into the air. Nearer to us than this recess could be seen a gentle declivity, forming a very shallow valley, in which once ran the Kaiwaka Stream, the former outlet to Lake Rotomahana. This once deep gully is now nearly filled to its top with ejected matter, to a depth of 80 feet, of stone, sand, and mud. All around this part of the crater edge the ground was cracked and fissured by earthquakes, and by the torrents of water ejected from the crater. Lying immediately to the west of it was a large deposit of mud, which extended some way up the range that divides Rotomahana from the Wairoa Stream, and on its surface were occasional pools of water, the remains of deluges cast out from the crater.

From this same spot a good view of the whole of the south end and top of Tarawera is obtained. The eye is immediately attracted by the altered appearance of the south-west end of the mountain.

Here a great rift—an enormous chasm—extends from the plateau-like top to the base of the mountain, ending (apparently) quite close to the site of the former Rotomakariri Lake. Various estimates have been formed of the dimensions of this great rift, and we believe that we are quite within the mark in stating it to be over a mile long, 500 feet wide, and 500 feet deep. No one, up to the present time, has been able to see the actual bottom of it. Out of this chasm rise, at several points, columns of dense black or brown smoke, not continuously, but intermittently; but no sign of any ejection of solid material was visible at the time. The edges were quite sharp and ragged, as if the solid rock had been ripped open by the enormous force of imprisoned steam; and in its upper part the ashes, rocks, and the ground generally for a long distance on either side, were coloured a yellowish-green, due no doubt to some of the products of volcanic action—such as ferric chloride. The slopes of the mountain around were covered deeply by ashes and stones, and near the base of it steam escaped from several cracks. As we sat on the surface of the sand observing the chasm through the glass, frequent shocks of earthquake caused cracks to open near the rift, and steam was seen to escape in little jets, ceasing, however, soon afterwards, as the cracks closed in or the loose materials fell into and stopped the vents. The southern end of the rift seems to be continued as a hollow right into the site of Rotomakariri, which is now occupied by a crater, from which rises a vast column of steam and occasionally smoke; indeed, this part seems to be one of the most active craters of the whole series. …

Riding home, weary and covered with mud, we halted to gaze upon one of the most glorious sights man could view. We stood in a light-timbered grove just outside the belt of the ash-covered plain, the setting sun at our back. Away and away in our front for miles lay the scene that not long since looked like snow, but now, reflected on it, the rays of the setting sun gave it the aspect of red coral. But, above all, there rose in solemn grandeur the towering mass of steam—thousands upon thousands of feet it ascended, until its crown was lost in the bright, fleecy clouds that came rolling up from the south. Bright, aye bright with the full effulgence of the orb which was still high above the horizon there; but lower, the dazzling brightness waned, and a faint glint of a golden hue was seen, to be rivalled by the richer colours and deeper gold of the nether parts until they deepened and sank through rose to carmine, and deeper hues suffused the base and the far-reaching crimson plain, while the deep greens of the bush in which we stood made up a picture difficult to equal, impossible to excel. And thus from earth to sky rolled the ever-changing mass of steam, rent at the base with the uprush of countless geysers, imparting to it changing and varying tints, beautiful and transient; but above, calm, solemn, and gorgeous, and apparently immovable. Slowly the deeper tints crept up, and left the base white and beautiful in the light of the bright full moon, while the crown still reflected the deep soft tints of a sun which had long since set with us.

From 'Observations on the Eruption of Mount Tarawera, Bay of Plenty, New Zealand, 10th June, 1886' by J.A. Pond and S. Percy Smith: *Transactions and Proceedings of the New Zealand Institute*, 1886.

The Insect-hunter

Between 1892 and 1951, a series of books by entomologist
George Hudson (1867–1946) were published, in which he
catalogued and described New Zealand's moths, butterflies,
beetles and spiders. While some of the books were scholarly
and systematic, others were popular works aimed at the
amateur enthusiast. In this excerpt from his first book, *An
Elementary Manual of New Zealand Entomology*, Hudson
provides a guide to collecting New Zealand's native insects.
He may sound like an old hand, but he is just nineteen years
old and has been in New Zealand five years.

Coleoptera, or beetles, may be found almost everywhere. Over-
turning logs and stones, peeling off bark, and cutting into the
solid wood of trees, all produce a great variety of species. A small
axe and an iron wrench, shaped something like a chisel, but bent
round at the upper end, are the best instruments for working old
trees. The bark should be all stripped off and examined, as well as
the surface of the log underneath. The same remarks apply to
stones, which should be searched as well as the places from which
they were removed. Sacks, if left about the fields for a few weeks,
often harbour good beetles, and when found they should always
be pulled up and examined.

An umbrella, held upside down under flowering shrubs in the
forest, will often be found swarming with beetles after the plants
have been sharply tapped with a stout walking stick. The same
object may be attained by spreading a newspaper, or sheet, under
the trees and then shaking them; the beetles will fall on to the
sheet, and may then be captured. The only advantage of the
umbrella is that it can be more readily used in awkward places,
such as on steep hill sides.

The dead bodies of birds and animals also contain peculiar species; they may be held over the umbrella and shaken into it, when the inhabitants will fall out, and can easily be obtained. Dead fish on the sea beach are often very productive. Moss and fungi are unfailing resorts of many of the smaller species of Coleoptera, and can be examined in the winter when the entomologist is otherwise idle.

Beetles should always be brought home alive. The small round tin boxes sold with Bryant and May's wax matches will be found very serviceable for this purpose. These boxes are far better for all kinds of collecting than either pill- or chip-boxes, as they do not break when knocked about. A separate box should always be given to a large or rare species, but most of the smaller kinds will travel quite safely in company, especially if a wisp of grass or a leaf is put into the box to give them foothold.

Beetles must be killed with boiling water, and left immersed some hours before setting. They must be pinned through either the right or left elytron,[*] and each collector must always keep to one side, as nothing looks worse than to see some of the specimens pinned, on the right and others on the left side. When pinned, the beetles are set on a corked board, the legs etc. being placed in a natural position, and retained until dry by means of pins and pieces of paper and card. The smaller species should be mounted with transparent gum on a neat piece of card, which can be pinned in the store-box or cabinet with the others. The greatest care should be taken to set symmetrically, so that the limbs on the right-hand side of an insect are in the same position as those on the left.

Hymenoptera[†] may be captured with the ordinary butterfly net, and are found abundantly during the summer. The larger

---

[*] A beetle's hard outer wingcase.
[†] An order of insects that includes wasps, bees and ants.

species are pinned through the centre of the thorax, and set in the same way as Coleoptera, the smaller ones on card with gum. These insects should, if possible, be made to fly into the vessel of boiling water, as by this means they generally die with their wings expanded, which is a great assistance when setting them. This can usually be managed by holding the box containing the specimen immediately over the water, and giving it a sharp tap with the finger of the other hand.

Diptera[*] are also captured with the net, and pinned in the same way, but should be killed with the laurel bottle.

Lepidoptera[†] are the most difficult of all to collect, and are at the same time the most attractive to beginners. They may be captured with a net made of fine gauze (mosquito net dyed green is the best material); the frame to support the net is constructed of a piece of cane bent into a hoop, each of the ends being supported in a forked tube shaped like a **Y**, and the long tube, forming the base of the **Y**, is firmly fitted on to the end of a walking-stick. This form of net is light, strong, and easily made; the only thing requiring special attention is the **Y**, but this can be readily made by any tinsmith out of two pieces of gas-pipe of different sizes, the larger one for the stick, and the smaller one for the ends of the cane to fit into. The collector should also be furnished with a number of small tin boxes. All this apparatus can easily be packed into an ordinary satchel.

When the entomologist reaches his hunting-ground, he will mount his net and place a number of the boxes in his left-hand coat pocket. The foliage of all trees and shrubs should be vigorously beaten and the insects captured as they fly out. When a moth is taken, the collector will first turn the net halfway round so as to close the entrance, and then, directly the insect ceases

---

[*] An order of insects that includes flies, mosquitoes and midges.
[†] An order of insects that includes butterflies and moths.

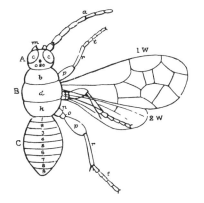

Body of an insect (Hymenoptera), showing the principal divisions: A, head; B, thorax; C, abdomen; *a,* antenna; *c,* compound eyes; *m,* mandible; *s,* simple eyes; *b,* prothorax; *d,* mesothorax; *k,* metathorax; 1W, forewing; 2W, hindwing; *n,* coxa; *o,* trochanter; *p,* femur; *r,* tibia; *t,* tarsus; 1 to 9, segments of the abdomen.

fluttering, he should carefully place one of the little boxes over it and slip on the lid. The box is then transferred to the right-hand pocket. He will soon learn to do this without in any way damaging the insect. On arrival at home, the insects should be immediately killed in the laurel bottle. This is an ordinary wide-necked bottle with a small bag of well-bruised *young* laurel shoots at the bottom, covered with a circular piece of card fitting accurately to the sides of the bottle. Laurel shoots can always be obtained about the middle of October, when several killing bottles can be prepared. They must always be wiped out before using, and kept carefully corked. After a few hours the insects should be tilted out of the bottle on to a tablecloth, and pinned exactly through the centre of the thorax. The rough surface of the tablecloth prevents them from slipping during the operation. About one-third of an inch of pin should project below the body of the insect. If a moth or butterfly dies with its wings folded upwards over the back, it must be carefully picked up between the thumb and index finger of the left hand, and the pin inserted with the corresponding fingers of the right hand. When all are pinned they should be transferred to a tin box, lined with cork, which has been previously well damped with water. While pinning them into this box great care must be taken not to allow the wings to come in contact with the damp cork. In about twenty-four hours the specimens thus treated will be ready for setting. This process is performed by means of corked boards of various widths for different sized species. Each board has a groove down the centre for the bodies of the insects to rest in, while the wings are spread out on either side. They should be carefully moved forwards with a fine-pointed needle to the desired position, and retained by strips of tracing cloth pinned firmly down at the ends. These strips must not be removed until the insects are thoroughly dry and ready to place in the store-box or cabinet. In setting Lepidoptera, as with other insects, symmetry and a natural position are the main points to be aimed at, special care being taken that the antennae, fore- and hind-legs, and wings, are shown in correct positions, the middle pair of legs being of course, in the majority of cases, hidden by the wings. It is almost needless to say that different sized pins should be used for various insects, but this point must be left to the discretion of the collector. Entomological pins of all sizes can be obtained from James Gardner, of 29 Oxford Street, London. Gilt pins are useful for many species which are liable to form verdigris on the pins, and are universally employed by many entomologists, but are probably not so strong as the silvered ones.

Many species of moths are only to be found at night. When working at this time the collector must suspend a bull's-eye lantern round his neck or waist, and can then have both arms free for capturing insects on the wing or at blossoms. Honey mixed with a little rum, and applied with a small brush to the trunks of trees a few minutes after sunset, will, on some evenings, attract large numbers of valuable species, but not infrequently it is quite unproductive. This mode of collecting has been termed 'sugaring' by entomologists, and may be employed during the whole summer. The best blossoms for attracting insects in New Zealand

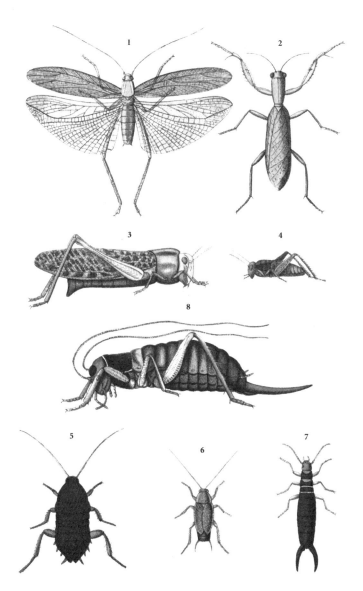

Hudson's beautiful illustrations were a feature of his books. On this page are: 1. katydid; 2. New Zealand praying mantis; 3. migratory locust; 4. New Zealand grasshopper; 5. black cockroach; 6. winged bush cockroach; 7. seashore earwig; 8. Wellington tree weta.

are those of the white rata, which blooms in the forest from February till April, and from which the collector may generally rely on getting a rich harvest. The insects can usually be slipped directly from the flowers into the killing bottle.

This is much better than netting them, although occasionally one will escape during the process. When dead the specimens should be placed in a small tin box which has been filled with cotton-wool, packed very lightly. In this way a large number of moths may be carried a long distance with perfect safety, and the extremely inconvenient process of pinning them in the field obviated. If Jahncke's patent boxes are employed it is quite unnecessary to kill the moths in the field. They can be boxed directly from the blossoms and taken home alive without suffering any injury.

Lepidoptera, and in fact all insects, are attracted by light, and in some situations the collector will find that he may frequently obtain good species by merely opening his sitting-room window and waiting for the insects to arrive. Much of course depends on the situation of the collector's residence and the nature of the night, which should be dark and warm. I have occasionally tried taking a lamp into the forest to attract insects, but have not met with much success. In swampy and flat situations, no doubt, attracting by light would be very effective, especially if a powerful lamp was employed, in an exposed situation, with a sheet behind it, supported between two poles. This method has been followed with great success by many English entomologists in the fens, but has not yet been tried in the New Zealand swamps, where it would probably be the means of bringing many new and interesting species to 'light'.

From *An Elementary Manual of New Zealand Entomology: Being an Introduction to the Study of Our Native Insects* by G.V. Hudson: West, Newman & Co., London, 1892.

Rutherford and the Discovery of Radioactivity

New Zealander Ernest Rutherford (1871–1937) arrived at Cambridge University's Cavendish Laboratory in 1895 as a student under the professor of experimental physics, J. J. Thomson. It was an exciting time for the young physicist to be in Europe: following Wilhelm Röntgen's sensational discovery of X-rays in November that year, Henri Becquerel would discover the spontaneous emission of X-rays by uranium-bearing compounds, and Marie Curie would discover two new radioactive elements, polonium and radium. In 1897 Thomson himself discovered the electron. In this account by Ernest Rutherford, prepared for publication from a lecture he delivered in 1936—he had died before he was able to provide a written version—the Nobel Prize-winning physicist outlines his role in the dramatic story of radioactivity.

When I entered the Cavendish Laboratory in 1895 I began work on the ionization of gases by X-rays. After reading the paper by Becquerel, I was curious to know whether the ions produced by the radiation from uranium were of the same nature as those produced by X-rays, and in particular I was interested because Becquerel thought that his radiation was somehow intermediate between light and X-rays. I therefore proceeded to make a systematic examination of the radiation and I found that it was of two types, one which produced intense ionization and which was absorbed in a few centimetres of air, and the other which produced less intense ionization but was more penetrating. I called these alpha rays and beta rays respectively, and when, in 1898, Villard discovered a still more penetrating type of radiation he called it gamma radiation.

In 1898 I went to McGill University, Montreal, and there I met R.B. Owens, the new professor of electrical engineering, who had arrived at the same time as myself. Owens had a scholarship which required him to do some physical research, and he asked me whether I could suggest a problem which he might investigate to justify this scholarship. I suggested that he might become familiar with the use of an electroscope by studying thorium, the radioactivity of which had in the meantime been discovered by Schmidt. I assisted him with his experiments and we found some very queer effects. It appeared that the radioactive effect of thorium oxide would pass through dozens of sheets of paper put over the oxide but that it was stopped by the thinnest sheet of mica, as though something was being emitted which could diffuse through the pores of the paper. The fact that the apparatus was very sensitive to the effects of draughts supported this diffusion idea. We next did experiments in which air was drawn over the thorium oxide and then into an ionization chamber, and these showed that the activity could be transferred with the air. However, if the air current was stopped, the activity in the ionization chamber did not cease at once but gradually died away in an exponential manner. I gave the name of 'thorium emanation' to this gaseous substance which could diffuse through paper, and could be carried away with the air and which preserved its activity for some time, decaying with a characteristic law.

I found that this emanation had the most peculiar property that when it was passed over bodies it made them radioactive. This appeared to be due to the deposit of a material substance, rather than to any activity induced in the bodies themselves under the action of the radiation, since the amount of the material deposited could be increased by applying an electric field. Many people at this time were obtaining capricious and peculiar results from materials placed near radioactive substances, and it seemed that these could probably all be explained by the presence of emanations of the type we had found in connection with thorium.

Before this explanation could be shown to be correct it was necessary to discover the exact nature of the emanation. This was very difficult, because the amount available was always very small. Soddy* and I concluded, early on, that it must be one of the inert gases like helium, neon, and argon, since it was never possible to make it combine with any chemical substance. We were able to make a rough estimate of its molecular weight by comparing its rate of diffusion with that of other gases with known molecular weights. By using the property of discharging an electroscope as a measure of the amount of emanation present, we were able to measure these diffusion rates with very small quantities of emanation. We concluded that the atomic weight must be of the order of 100. We next tried to find whether the emanation was produced directly from the thorium, or from some intermediate product. Using chemical methods we were able to separate an intermediate substance, which we called thorium X, from which the emanation was produced.

About this time Ramsay showed that helium was present in most radioactive minerals, and that it represented another gaseous product of the transformations. Later on I was able to show that the helium was due to the accumulated alpha particles.

Radium was not available in any quantity till 1903 or 1904, and most of what there was in the world was in the possession of the Curies, who had separated it by a long and arduous process from pitchblende. One of the first observations they made was that a quantity of radium weighing about 100 milligrams kept itself above the temperature of the surrounding air, and they deduced

---

* Frederick Soddy, a chemist at McGill University in Montreal, who worked with Rutherford on radioactivity.

that a gram of radium would emit heat at the rate of 100 calories per hour. This experiment created great excitement, because the idea of any substance keeping itself permanently at a temperature higher than its surroundings was repugnant to the old-fashioned physicists, and the prevailing idea became common that radium had a peculiar property of acting as a thermodynamic engine using heat from the air. I was firmly of the impression that the heating effect was a necessary consequence of the emission of the alpha and beta particles and that it decreased with time in the same way as the activity. Later on we were able to classify the heating effects of radioactive bodies and to show that there was nothing obscure about the process. We were able to show that heat can be evolved in enormous quantities in these radioactive changes; when reckoned per unit mass of the material these quantities are millions of times greater than those given by chemical reactions, and we were able to show that this is a characteristic of all radioactive changes. ...

By various experiments and with the help of various collaborators I was able to show, by deflecting alpha particles in magnetic fields, that these particles were helium atoms carrying two positive charges, and we were also able to determine their speed.

From 'The History of Radioactivity' by Ernest Rutherford in *Background to Modern Science*, edited by Joseph Needham and Walter Pagel: Cambridge University Press, Cambridge, 1938.

A Most Incredible Event

Ernest Marsden (1889–1970) came to New Zealand from England in 1915 to replace Thomas Laby as professor of physics at Victoria University College in Wellington. In 1922 he turned from research to bureaucracy, working first as assistant director of education and then, from 1926 to 1954, as head of the Department of Scientific and Industrial Research. In 1909, while a student of Ernest Rutherford's at Manchester University, Marsden had been fortunate enough to play a crucial role in Rutherford's development of the nuclear model of the atom, through an experiment now known as the gold foil experiment. In his 1948 Rutherford Memorial Lecture, Marsden recalled the experience.

I was then a callow youth in the second year of my honours course. The arrival of Rutherford decided my fate—and what a spur to energy he was! I quickly ran through all the practical exercises in electricity and magnetism and optics etc., and then proceeded as the first of the guinea pigs in those beautiful elementary experiments in radioactivity drawn up by Rutherford and Geiger, and afterwards published by Makower and Geiger. We constructed our own electroscopes and obtained first-hand acquaintance with the absorption and other properties of the alpha, beta and gamma rays, and with the radioactive transformations of radium emanation and its products—and similarly with thorium and actinium.

Soon I approached the frontier of new knowledge, it did not take so long in those days, and I was allowed to help Rutherford in experiments on phosphorescence produced by alpha and beta rays, and later to act as Geiger's assistant in experiments on the scattering of alpha particles, compound scattering as we later termed it, and on the counting of alpha particles using the method, then under development, of magnification of the effect through ionisation by collision and using a Dolezaleck electrometer. Geiger was a master of this instrument, which needed much skill if accurate results were to be obtained. I was a member of the O.T.C.* and one Saturday morning, when working in uniform preparatory to an afternoon's military exercise, I happened in sheer admiration to say, 'Dr Geiger you *do* know your electrometer.' His reply was, 'Ah, but I know my gun better than I know my electrometer.' Although he was a Frank and not a Prussian, such was the effect of conscription. Alas, Geiger was to fight against us as an artillery officer in the war which followed five years later.

Geiger was a delightful man, energetic, orderly, kindly and methodical, and he was most loyal to the Professor. We tried to teach each other our respective languages. He read Eddington to me and I read Planck's *Thermodynamics* to him and we tried to correct each other. I hope I did not give him too much of a Lancashire accent.

Rutherford was also engaged with Royds in that beautiful experiment on the spectrum of the radium emanation and production of helium from alpha particles fired through a thin-walled container of radium emanation, which showed definitely that the alpha particles were helium. I remember the details so well because the spectroscope was set up in the same room as that in which I worked on phosphorescence. Photographs of the spectrum of radium emanation and helium were taken at intervals as the emanation decayed and the helium spectrum appeared. One day someone had been too inquisitive in looking at the spectroscope. Rutherford came into the room and, noticing that the prism had been displaced, flew into a towering rage. He came over to the

---

* Officers Training Corps.

optical bench at which I was working and placed his hand round the back of my neck, only moderately gently, and said, 'Did you move that prism?' I knew and trusted him too well to have the slightest fear of him and answered, 'No.' I was sufficient of a schoolboy to enjoy the 'boss' in a tantrum. Half an hour afterwards he came back to the room and sat deliberately on a stool alongside me and quietly expressed his apology for getting his 'dander' up and accusing me of the misdemeanour. He must have found out the culprit in the meantime. Then we went on to one of those helpful and, on his part, non-condescending discussions of the progress of my work, which all who had a similar privilege treasure as their happiest recollections. I have never known him to bear malice. You will appreciate his action when I add that I was only nineteen years of age at the time—in fact I realise how extraordinarily fortunate I was to work under his aegis and to make two contributions to the Royal Society, and be allowed to read one of them myself, before I had passed the age of twenty.

One day when I was exercising my privilege of working with Geiger, Rutherford came in and a discussion ensued between them as to the nature of the huge electric or magnetic forces which could turn aside or scatter a beam of alpha particles when passing through such a thin film of gold (some 10^9 gauss). Rutherford turned to me and said, 'What about trying whether you can get alpha particles reflected from a solid metal surface.' I don't think he expected any such result, but it was one of those 'hunches' that perhaps some effect might be observed, and that in any case that neighbouring territory might be explored by reconnaissance. Rutherford was ever ready to meet the unexpected and exploit it, where favourable, but he also knew when to stop on such excursions. Naturally, I knew enough to appreciate that even though a negative effect might be expected yet if I missed any possible result it would be an unforgivable sin. Accordingly, I made quite sure of preparing as strong a source of radium emanation as I possibly could in a conical tube closed by a mica window—a technique which had been developed by Geiger. To my surprise I was able to observe the effect looked for…

> Under the then prevailing 'plum pudding' model of the atom, in which negatively charged electrons (the 'plums') were distributed throughout the positively charged atom (the 'pudding'), the alpha particles were expected to pass directly through the thin gold foil. In a 1936 lecture, Ernest Rutherford recalled his astonishment at Marsden's report that some of the alpha particles had been deflected through large angles, or even reflected back towards the source.

It was quite the most incredible event that has ever happened to me in my life. It was almost as incredible as if you fired a 15-inch shell at a piece of tissue paper and it came back and hit you. On consideration I realised that this scattering backwards must be the result of a single collision, and when I made calculations I saw that it was impossible to get anything of that order of magnitude unless you took a system in which the greater part of the mass of the atom was concentrated in a minute nucleus. It was then that I had the idea of an atom with a minute massive centre carrying a charge. I worked out mathematically what laws the scattering should obey, and I found that the number of particles scattered through a given angle should be proportional to the thickness of the scattering foil, the square of the nuclear charge, and inversely proportional to the fourth power of the velocity. These deductions were later verified by Geiger and Marsden in a series of beautiful experiments.

> Rutherford pondered the result for two years, and in 1911 he proposed an atom with a centralised concentration of mass

and positive charge—which he later called the nucleus—surrounded by empty space and a sea of orbiting, negatively charged electrons. In 1913, the Danish physicist Niels Bohr extended Rutherford's model of the atom by fixing the energy levels in which electrons could orbit the nucleus. Bohr explained that atoms emitted fixed amounts (quanta) of energy, or radiation, when electrons jumped from one stable orbit to another. The 'Rutherford-Bohr' model became the accepted model of the atom, and is essentially the picture of the atom we use today.

From 'Rutherford Memorial Lecture, 1948' by Ernest Marsden; and 'The Development of the Theory of Atomic Structure' by Ernest Rutherford, in *Background to Modern Science*, edited by Joseph Needham and Walter Pagel: Cambridge University Press, Cambridge, 1938.

Booming Kakapo

Richard Henry (1845–1929) was the caretaker of Fiordland's Resolution Island, a sanctuary for flightless birds,
from 1894 to 1908. In 1903 he published *The Habits of the
Flightless Birds of New Zealand,* in which he warned of the
possibility of kakapo becoming extinct. In this extract, he
writes of his observations of and interactions with kakapo,
and postulates on the reasons for the bird's intermittent
breeding habits—they seemed to breed only every second
year. It is now believed that kakapo breeding is linked to the
availability of key food supplies, such as the fruiting of
rimu trees.

The kakapos of Resolution Island eventually fell victim
to stoats, but other kakapo populations survived. Today
there are eighty-six known kakapo in New Zealand, and
they are managed under the Kakapo Recovery Programme.
One of the birds is known as 'Richard Henry'.

The great ground parrot of New Zealand is called 'kakapo' by the
Maori. I think it is the largest and the only one out of the great
family of parrots that cannot fly. Probably its ancestors could fly,
and, like the rails, came here on the wing; but the absence of
enemies on the ground, and the abundance of food, allowed the
muscles of the wings to degenerate and those of the legs to
develop, until now they are fairly good runners, and their wings
are only for ornament, or at most to prevent them being hurt
from a fall, for they love to climb about among rocks and trees in
search of berries and seeds.

There is evidence that a great hawk once lived in New Zealand,
and even now there is a fierce little hawk that delights in knocking
down birds on the wing, so that the kakapos' forefathers may
have been forced to give up flying, those alone surviving that took

shelter in the undergrowth. Its breastbone has just the trace of a keel, so that it must have taken a long time to alter its shape to what it is … And just now the 'lords of creation' have imported ferrets and weasels that prey on all such things that sleep on the ground, and, as kakapos cannot be expected to learn in a day what their race had forgotten for thousands of years, the chapter of their history is in all likelihood coming to a close. Fortunately they have many friends, and the New Zealand Government takes a kindly interest in their affairs, and has appointed two reserves and men to put them out upon islands with some of their help-less neighbours, where, if fortune favours, they may long survive.

Kakapos may be called night birds and fruit-eaters, which is an unusual combination of characters, but they also eat grass, leaves, and some fungi when the fruit is over. They chew their food more effectually than any other birds that I am acquainted with. For this purpose there are diagonal grooves in the upper mandible, in contact with which the lower acts in the manner of a steel mill. On examining the food in their crop it is found to be so well masticated it is impossible to tell what it is; yet by a wise provision some seeds are passed entire—such as tutu, which are poisonous, and mapou seeds, the berries of which are an important item of kakapo food. When they chew some of the fibrous grasses, reeds, or flax they leave the fibre in little pellets attached to the remainder of the leaf, so that they know how to use their simple milling apparatus. There is a disc of feathers around the eyes like those of the owl, and like other creatures that go in holes they have hairs about the nose, or feathers produced into long point-like hairs.

Some of them rest on the ground under ferns during the day, but the great majority prefer to take up their quarters in hollows and dens in the moss among the roots of trees, where they sit on a root in the gloom all day, and only come out in the dusk of the evening. So well does their colour accord with the yellow and green of the ferns that it is impossible to see them unless they move. Of this they are well aware, and often keep perfectly still even when within arm's length. They are simple, poor things that know nothing of enemies. Once when without a dog I met one sitting on a stick under a fern a few feet from the ground, and went up to have a talk with it. It looked at me more in wonder than fear, until I chucked it under the chin, when it assumed a fierce attitude and protested in its hoarse voice, but made no attempt to go away, and when I let it alone for a few moments it coolly put its head under its feathers and went to sleep again.

They have their family quarrels, of course, and sometimes scandalous fights, for I have found both males and females with their eyes seriously injured and old scars on their heads, and it is by no means a very rare thing to find a female with only one eye, for it is their misfortune to have powerful beaks and claws. I must never put two in one cage, for they seem to blame each other for their trouble, and start fighting at once.

The tail of the female is longer than that of the male, and she is greener in colour, with less yellow on the head and breast. She is also less in size, and seldom very fat like her lazy mate—if ever she has a mate in the ordinary sense of the term, for they are the most solitary of birds. She makes her nest on the ground in some of the mossy dens, and lays from two to four white eggs like those of the harrier hawk. I never found two birds in the one den at any season, though there is room for a dozen, and I think that the male never goes near a nest, and knows nothing about it.

They only breed every second year—not independently, but all breed one season and none the next—and it is a great puzzle to naturalists why some do not breed in the off season, or how they all come to such a unanimous agreement about it. Months before the appointed breeding season the male is developing an air sac in his throat, which he can puff up like a drum, and which may act like a sounding board to assist in making the curious drumming notes in the spring. This note is not unlike the boom of the bittern, but is repeated five or six times in succession, and can be heard at a great distance. It excites curiosity not easily satisfied, and has caused some discussion and difference of opinion. A surveyor of large experience denied that it was a kakapo at all, and asserted that it was the rare notornis; and a high authority laid it down that the air sac was just outside the windpipe, and therefore not connected with the voice; but it is not necessary for air to pass through a drum to make a sound. A whole party of bushmen set themselves the task of finding out what it was, and came to the conclusion that it was not a kakapo, for when they followed up the sound and got quite close to it at night there was a moment's silence, and then it began again half a mile away, so they were certain it could fly. But it was only another kakapo, while the first one may have been at their elbows.

I was twelve years on the dry side of the mountains, near Te Anau Lake, and had ample opportunities for observing their habits. I heard their drumming every alternate year until 1892, which was their due season, but they did not breed that year, and skipped two years in succession. Now, they must have held a meeting about the projected alteration in their programme, and a wonderful meeting it was, no doubt, regarding its decision, for not a drummer was heard that year. To realise the wonder of it we must remember that they had to come to a decision about six months previously, for the preparation of their drums; so that it could not have been the effect of the fruit and the flowers of that season.

This is my second year on the wet side of the mountains in Dusky Sound, and now I find that the seasons here and at Te Anau coincide, so that delegates were needed from both sides of the Southern Alps, and they all agreed.

I have exhausted all my speculations about flowers and fruit and physical conditions to account for this intermittent season, but all in vain, and generally have to fall back on the idea that they have useful social laws as mysterious as those of ants or bees. This idea would be acceptable if they lived in close communities, but it is difficult to understand when we know that they are solitary birds, living in a rough country so dense with undergrowth that it is all the time like getting through a hedge.

About 1886 ... I found about a score in a few hundred yards, and there may have been about a hundred within a mile—but only on one occasion since have I known a number to congregate in so small a space; so the idea of a meeting may not be altogether fanciful.

It appears as if the breeding season was controlled by the males, for when there is no drumming in the early summer there are no eggs or young ones. And if they willingly missed two years in succession what a vision of self-denial and forethought it opens up! But should it be the density of population, and consequently the supply of food during the previous winter, that influences their conduct, it is a very beautiful arrangement to save a waste of life and labour. This would also be the most acceptable theory if it were not for the fact that every valley on the wet and dry sides of the mountains could not be in the same position as to food, so that we have to come back again to 'social law'—even to the idea of a captain or a queen to adjudge population and order their conduct accordingly. ...

There are considerable areas in the bottoms of valleys and around landslips where berries are produced in great plenty, and such places are called 'kakapo gardens'. Then there may be miles of beech forest which apparently produce very little food; yet the birds wander away anywhere after the breeding season, even out on the grass on the mountain-tops. When the breeding season comes round again the males take up their places in the gardens, and about the 26th November commence their booming love-songs. These notes do not sound loud when close by, but can be heard many miles away, and may be useful in calling in the females to suitable places for nesting and feeding their young. Under favourable circumstances I have heard it at a distance of six miles, and somehow the humming, murmuring sound made the season appear more lonely and peaceful, giving a faint idea of how it might sound to their half-lost fellows away in the bush. In the virgin forest, where nothing heavier than themselves has ever trodden the yielding moss, they have very distinct pathways, especially going up some small hill, where in the driest place on top a gleam of sunshine may enter among the trees. Here will be several 'dusting-holes' and signs of traffic, as if many birds go up there occasionally, either singly or in companies. But this habit of only coming out at night enables them to keep nearly all their social affairs to themselves.

When newly hatched the young are covered with snow-white down, and they remain in the nest until nearly full-grown. During this period the mother's feathers are all draggled and worn, and I often wondered how she could tramp away and carry home sufficient food to keep two or three young ones like balls of fat. The males are also very fat, while the mother is like skin-and-bone, and once I thought that, phoenix-like, she might die when she reared a brood; but I soon found that idea untenable, though it would in a sort of way account for the intermittent breeding season.

When we are hunting for kakapos our fox-terrier is often at fault, running here and there in an excited manner, and finishing by finding nothing. It is a long time since we found one up a tree. In fact, we only got two up trees altogether, and I could not imagine kakapos running away from anything, for I always

John Buchanan (1819–98), a Scottish-born botanist and artist,
travelled widely through New Zealand, recording its flora, fauna
and landscapes in a collection of notebooks, sketches, paintings
and academic articles. In this wood engraving from around 1860,
Buchanan shows a pair of kakapo, the first scientific description
of which was made by George Gray of the British Museum in
1845. By the time of this 1860 sketch, kakapo were already extinct
on the North Island and declining in numbers on the South
Island. E-208-q-005-1, Alexander Turnbull Library, Wellington

thought that they knew nothing of enemies. However, I have often got them with one eye out, and with deformed nostrils, as if bitten or bruised. Recently we caught one with part of the skin scraped off the top of the head, and both eyes so severely injured that it was nearly blind. It could not have been a ferret that caused the injuries, because wekas were plentiful, so it must have been another kakapo.

One of the last days we were out our dog tracked one up a tall sloping stump, and hunted a light young kakapo off the very top. We saw it flutter down, but it was so artful in hiding that it was some time before we found it. Now, this was a plain case of a light, active young bird trying to avoid a well-known enemy; and now I am quite satisfied that the old ones—probably the old males—persecute the young ones, and perhaps kill them. We found one just dying, with hardly life enough left to attract the dog. I thought it was starved, and did not look for marks or injuries. This may be only a wise arrangement to prevent too close breeding, for I think it is quite common for many animals to fall out with their young ones when they grow up. At this season, when food is scarce, they are the most unsociable creatures living, for each one seems to have half a square mile to itself. …

On 26th December, 1897, we anchored in Cascade Harbour, and long before dark we heard the kakapos booming, after a silence of two years, and later on in the evening we heard them from all sides, though we had so often hunted there that we thought we had pretty well cleared them out of that place. Seven days previously we were under Mount Pender, where there are plenty of kakapos, but we heard no booming, so that they must have just commenced. At Te Anau they used to commence about the 1st of December, which date they kept to fairly well all the years I was there.

Next day we went hunting on the west side of Cascade, and caught three very fat kakapos. There were no berries or seeds on any of the trees, and we were puzzled to know what they got so fat on. There was a tiny seed or blossom on the carpet of green moss that covers that country, and I often saw little holes that they scraped out with their beaks, perhaps looking for truffles or fungi of some sort; but they chew their food so well in their milling beaks that I could not identify it. It was important to know what they were eating then, because it might solve the mystery as to their intermitting breeding seasons, but I could not find it out. Their crops contained mostly a green pulp, with some of a lighter colour, but what it was I could not tell.

The male can swell up his air sac (of which the female has no trace) nearly as big as his body, so that he must be a formidable-looking fellow on parade. I never saw one booming, however, for they never boom in captivity.

I took a special trip to Wet Jacket Arm to try and get better acquainted with these birds, and on the 21st January, 1898, climbed a high ridge south-west of Oke Island. It was very steep and rough, and all along its narrow top for half a mile were 'dusting-holes', as I used to call them, but there was not a particle of dust in them, as there had been about an inch of rain every day for a month. So 'dusting-hole' is, I think, therefore, a bad name; 'bower' would be more suitable. They were about 18 inches in diameter, fairly level on the bottom, and 3 inches deep, with steep sides. In some the peaty earth was pressed down firmly as if by the naked hand, while in others it was freshly raked up and loose. They were all connected by fresh, well-beaten pathways, so that a good many birds must go up there of an evening at this season, though in the off-season these places are deserted. This suggests that they are used for dancing or parades in their courtships. The Australian lyre-birds also make those holes, or ones like them.

Someone has suggested that the booming may be a defiance or challenge between the males, as in the case of cocks crowing; but I think that, owing to the thin population of kakapos in this dense forest, and the poor means of travelling, it was necessary for either the male or female to have a loud call. The voice of the female kakapo is a hoarse cough, and can only be heard for a couple of hundred yards, while the male's booming can be heard for a couple of miles. Therefore I think it likely that the males take up their places in these 'bowers,' distend their air sacs, and start their enchanting love songs; and that the females, like others of the sex, love the music and parade, and come up to see the show—that is, if they can see the green and yellow in the dark; if not, they can tramp along the pathways, listen to the music, and have a gossip with the best performers. However, it is almost certain that they can see distinctly, because the plumage of the male bird is pretty, and always looks its best at this season, and he would not retain that distinction without a reason for it.

Though we can hear plenty of kakapos in the evenings, we can never tell within a mile where they are, and they do not keep the booming going long enough for us to hunt them up. They start with a couple of short grunts, and then five or six deep measured notes like the sound of a muffled drum, the loudest in the middle. This series will be repeated about three times in the daylight, and then there will be silence until some others take up the cry, perhaps miles away.

On this ridge we got quite close to one when drumming, and it was a powerful note. I could feel the tremble of it, and my boy who was holding the dog 30 yards away could also feel it. I thought the drumming was just at my feet, and we stood still for a long time in hopes that the bird would commence again, but he was silent, and when we brought up the dog we found him 40 yards away, where he had taken shelter under a log. We had come up with all caution, stopping when he stopped, and walking while he was drumming, yet he seemed to have taken alarm. …

'Kakapo' is from two Maori words—*kaka*, a parrot, and *po*, night—which is very becoming, because I think they are the only parrots that feed at night. They have small eyes for night birds, and often climb trees in the daytime to sit in the sun after a spell of wet weather, which shows that it is not the light they fear; but probably, like many other creatures, they have chosen the night to feed the better to avoid their enemies. The only enemies they have here are the sandflies, which do not come out at night, but collect very quickly about any game they find near the ground in the daylight. The kakapos' slow movements would allow them to be punished very severely if they walked about on the ground in the daytime, for I know to my cost that the flies are expert at getting in under cuffs and collars, and may do the same with the kakapo's loose feathers. Therefore, when the sandflies have gone to bed the kakapos come out and gather food in peace, and retire to their dark places in the morning where the sandflies will not enter.

When I am in a penguin's cave here I can always see near the door a cloud of sandflies that will not enter even into the gloom where I can see quite well. A good many of the wiser penguins seem to know how far the flies will come in, but some of them have their nests too near the door, where their young ones will be punished severely, if not killed outright. I had two captive roas[*] killed by sandflies.

I was always puzzled to know what the kakapos got so fat on in summer-time, but now I find that they suck the honey out of the rata blossoms, like all the other bush birds, and as this honey is plentiful in the Sounds in December it is an important food to mix with their various other items. I went out on the 18th of

---

[*] Kiwi.

December and gathered a teaspoonful of this honey in ten minutes with a little glass syringe, so that the kakapos could get as much of it as they wanted. The little branches of the ratas are very strong and able to bear their weight.

From *The Habits of the Flightless Birds of New Zealand: With Notes on Other New Zealand Birds* by Richard Henry: Government Printer, Wellington, 1903.

Kakapo

The kakapo's strong and distinctive smell was one of the reasons for its decline: it was a giveaway to predators—dogs, cats and other introduced mammals—against whom this ground-dwelling bird had little defence. The world's largest parrot now exists only in carefully managed populations on offshore islands. This poem is by Auckland poet Sonja Yelich.

a kakapo is the
smell of some
honey,
papaya
& the inside of
an old clarinet
case
which is spitty
have a go kakapo
sniff yourself

From *Trout 10*: www.trout.auckland.ac.nz/journal/10/10_3.html.

A Trip to the Seaside

George Malcolm (called G.M.) Thomson (1848–1933) taught science at Otago Boys' High School and later entered politics as the member of parliament for Dunedin North. Like any gentleman naturalist worth his salt, he took ladies on nature rambles, kept a herbarium, participated in learned societies, and wrote and illustrated books and papers—in his case, on botany, crustacea and introduced wildlife. Thomson had an enduring interest in marine life, and one of his most significant achievements was the establishment of a marine research station at Portobello on the Otago Peninsula. In this excerpt from his 1909 book *A New Zealand Naturalist's Calendar and Notes by the Wayside*, he writes of the delights of the New Zealand seaside.

A seaside place would be of little interest if one could not go at low water and poke among the rock pools. The reefs on the ocean face of Moeraki Headland form an excellent hunting ground for the naturalist, as they run out for a considerable distance, exposing at low water long flat shelves of seaweed-covered rocks full of large and small pools. Such places are a never-failing source of interest. After a storm, when the waves have torn off great masses of tangled weed and sea-wrack and hurled them up with all their living freight on to the sandy beaches above the reefs, the collector can get a harvest of material. But if he comes on the beach some days after a gale he will find how rapidly a heap of seaweed is attacked and removed once it is out of the reach of the waves. If a frond be seized and dragged away so as to open up a heap, myriads of flies are exposed all busy feeding on the decaying weed. The number of these flies is marvellous; they seem to form living masses in suitable localities. If we drag away the weed till the sand is bare we meet with crowds of shore-hoppers, which skip away briskly or burrow rapidly into the sand. Among these may be found large black earwigs, some of the curious staphylinid beetles with their short wing-cases which only cover half their bodies, and also small black ants. The flies and the hoppers mainly feed on the weed, though the latter are as thorough going cannibals as can be met with, while the earwigs and the ants are carnivorous and help themselves to the other scavengers.

But the pools are the chief source of attraction. They are seen at their best after a few days of quiet weather, when the various animals which live in the shallow coastal zones of the sea are tempted to come shorewards without the risk of being pounded to death by the unceasing pitiless surge. Shrimps and prawns of several kinds, hermit-crabs, sea slugs, and other curious denizens of the deep, creep up at high water on to the weed which covers the reefs, and as the tide recedes they are left, along with various species of shore fishes, in the pools. It is interesting to notice how much protective colouring comes into play among these creatures. The fishes of the rock pools are usually blackish or dark gray with whitish stripes or markings, so that when they lie still in the weed-fringed clefts of the rocks they may be looked straight at without being seen. It is only when they move that their presence is revealed, and then with a dart they dash through the limpid water for a foot or two and again become invisible. The shrimps and worms which are exposed are frequently of a greenish hue so as to harmonise with the prevalent weed. There is a common red shrimp also found on these reefs—a peculiar little species which progresses by active leaps when disturbed, but this always takes care to hide under stones. Similarly, the reddish worms, which are always snapped up quickly by fishes when exposed, dwell mostly in tubes, and only put their heads out to see what is going on in the outside world. The sea slugs are usually very brightly coloured, and are really most elegant creatures when seen in the pools, with their pearly white or clear lemon-yellow bodies and their beautifully plumose gills. They are so conspicuous that we must suppose that they are not very good eating, otherwise they would be quickly decimated. Sea anemones of many hues and sizes often line the cracks between the rocks in the pools, opening out their wide tentacles like the petals of a cactus. If we drop a worm or a bit of meat into the centre of the flower the arms close rapidly and engulf the morsel.

One reason why pools on a rock-bound coast are so interesting is no doubt the constant element of change which they manifest. Not only are the objects very numerous, and often very peculiar, but there are constantly new forms appearing, and the variety is infinite. It is this unceasing changefulness of aspect which makes the sea itself so interesting, and which keeps one from wearying when contemplating its protean moods.

From *A New Zealand Naturalist's Calendar and Notes by the Wayside* by George M. Thomson: R. J. Stark and Co., Dunedin, 1909.

The Much-maligned Mangrove

Leonard Cockayne (1855–1934) was a pioneering New Zealand botanist who worked during the transition between the collecting and cataloguing era of botany and the move to a more holistic study of plants and their environment. In this excerpt from a chapter about coastal plants in his book *New Zealand Plants and Their Story*, Cockayne writes fondly of the mangrove, a small tree that populates intertidal areas in the north of New Zealand.

Now, quite undeservedly, the mangrove has got a bad reputation. A mangrove swamp is supposed to represent all that is most hideous on earth—alligators in crowds, a fearsome odour, crabs waiting to pick such of the victim's bones as are left by the alligators, malaria, and deadly 'microbes' in vast abundance. Even in the tropics this picture has been shown to be absurd, but in New Zealand the mangrove belt is quite a pleasing feature of the northern rivers. The mangrove is also a beneficial plant, as it materially assists in turning muddy useless shores into good dry land.

Moreover, the mangrove is one of the most noteworthy plants in nature. As our boat proceeds up the river the tide has turned, and the slimy flats where the mangrove is rooted come into view. There, projecting out of the mud, are thousands of upright bodies, six inches or so in height, looking much like stout asparagus-shoots … One might feel sure these were young mangroves; but they are nothing of the sort, strange as it may seem: they are, in fact, roots which, instead of passing downwards to anchor the trees, grow upwards into the air. On being examined, these erect roots are found to consist largely of a very porous tissue. Plants, like animals, cannot live without oxygen. They need to breathe just as much as we human beings do; without air they would die of suffocation, nor would they get energy to carry on their work. In the soft mud is little of the life-giving gas, hence the necessity for the mangrove to obtain a supply for its ordinary roots. This it does with these erect organs, which are the veritable lungs of the tree. Of course, the aerial parts of the mangrove, like those of any other tree, procure oxygen by means of the small pores in the leaves and minute openings in their bark.

The mangrove, too, has another peculiarity of even greater interest than that just described. If a seed were to fall on the muddy floor of a tidal estuary, being washed hither and thither by the ebb and flow of the tide, it would have little chance of germinating. While still on the tree, however, the seed has germinated. The little mangrove (embryo) has emerged from its seed-coat, awoke from its heavy sleep, and become a fair-sized seedling plant with rudimentary roots. At this stage it somewhat resembles a broad bean which has just germinated. Therefore it is not a seed which falls from the tree, but a young growing plant, whose roots rapidly increase in length, pass downwards and outwards from near the tip of the stem below the seed-leaves, and anchor the plantlet firmly in the unstable ground. The green seed-leaves, with which the little plant was also provided while attached to the tree, can manufacture food-material; but this is not all, for they are also fleshy and full of nourishment, so the young mangrove can live until the time when, provided with foliage, it is in a position to manufacture for itself in sufficient quantity the sugary foods it requires from the carbonic acid of the atmosphere and water. Surely none need cast contumely on such a plant as this!

From *New Zealand Plants and Their Story* by Leonard Cockayne, second edition: Government Printer, Wellington, 1919.

From Sea Floor to Coastal Plain

Charles Cotton (1885–1970), professor of geology at Victoria University College from 1921 to 1953, was the author of six internationally acclaimed books on geology and geomorphology. With their simple text and clear illustrations, Cotton's books were widely used as geography textbooks, but were just as likely to be found on the bookshelf of a New Zealand farmer. Artists, including the painter Colin McCahon, were also inspired by Cotton's drawings.

Cotton was influenced by an American geomorphologist, William Morris Davis, and followed his approach of illustrating the evolution of landforms with a sequence of diagrams that showed the landscape's progression from youth, through maturity, to old age. This extract and the accompanying illustrations are from *Geomorphology of New Zealand*, Cotton's first, and most influential, book.

Coastal Plains—When a portion of the sea floor emerges to become land, the uplifted portion is commonly a strip, narrow or broad, termed a *coastal plain,* bordering a pre-existing land (*old land*), which has been uplifted along with it. The uplift of a coastal plain may or may not be accompanied by deformation. A coastal plain of simple structure—that is, uplifted without notable deformation, though perhaps gently tilted seaward—serves as an example in connection with which may be considered the dissection of a flat area with very little slope.

Such a coastal plain is shown diagrammatically in fig. 66. Block *B* shows the initial form exposed by withdrawal of the sea.

The majority of the rivers on a newly emerged coastal plain are the rivers of the old land extended across the newly uplifted sea floor and seeking the sea by the easiest (consequent) paths. These are *extended* rivers. In the simplest case their courses are straight,

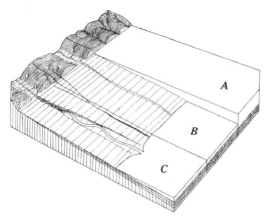

Fig. 66 — Diagram of a coastal plain of simple structure.
Block *A* shows the old land before uplift, block *B* the newly
emergent coastal plain, and block *C* the same after extended
rivers have become graded in the soft coastal-plain sediments.

parallel with one another, and at right angles to the shoreline;
but, obviously, even small irregularities of the initial surface will
cause the rivers to take less direct courses, and two or more may
unite before reaching the sea. These extended rivers, carrying as
they do a considerable volume of water when they leave the old
land, are competent to cut down and grade their courses quickly
in the weak sedimentary rocks of the coastal plain. Broad areas of
the flat interfluves may, however, long remain undissected.

In New Zealand, there is a strip of coastal plain of simple
structure bordering western Wellington and part of Taranaki to a
width of several miles. It was exposed as a result of uplift of about
600 feet … and initially extended farther seaward than it now
does, for it is bordered at the margin by wave-cut cliffs. It is still
young, but is crossed by the deep, mature valleys of extended
rivers. The broad belt of maturely or sub-maturely dissected weak
rocks farther inland has also been described as a coastal plain,
but, as the position of the old land relative to it is not certainly

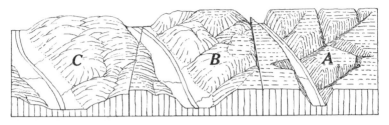

Fig. 67 — Diagram illustrating the dissection of an uplifted plain by insequent branching streams. *A*, young stage; *B*, dissection approaching maturity; *C*, mature stage.

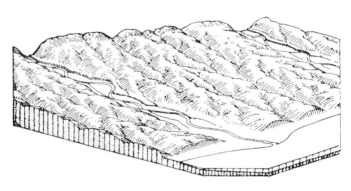

Fig. 68 — A coastal plain maturely dissected by extended consequent, new consequent, and insequent streams.

known, it is best regarded simply as an uplifted portion of the sea-floor. The geomorphology of this area has not yet been worked out. …

Insequent Streams—At a somewhat later stage new tributaries are developed. These start as steep ravines cut by concentrated rain-wash collecting in slight hollows accidentally formed in the steep sides of main valleys. As these gullies grow longer and deeper they receive an increasing amount of water both as surface run-off and as seepage through their steep banks. They rapidly eat their way back into the interfluves by *headward erosion*. Streams starting in this way, the positions and directions of which

Fig. 69 — Insequent drainage pattern exhibited by the tributaries of the Wanganui [Whanganui] and neighbouring rivers.

are purely accidental except in so far as they are determined by the slopes of the sides of the main valleys into which they flow, are termed *insequent* (figs 67, 68). They in their turn develop insequent tributaries, which also work back headward into the interfluves, so that the area of the undissected surface is reduced with increasing rapidity.

The pattern, as seen on a map, which is developed by insequent drainage has been likened to the branching of an apple tree, and has been termed dentritic (fig. 69).

From *Geomorphology of New Zealand* by Charles Cotton: Dominion Museum, Wellington, 1922.

143

Shell Collecting in the 1920s

Charles Fleming (1916–87) was one of New Zealand's last great naturalists. An expert in geology, ornithology, entomology, conchology, palaeontology and biogeography, and an advocate for conservation of New Zealand's forests and birds, he was knighted for his work in 1977. Like many of his scientific peers, Fleming's interest in the natural world began when he was a child. In this piece written in 1981, Fleming recalls his youthful interest in shell-collecting. A.W. Baden Powell, whom Fleming meets on one of his beach excursions, was in 1929 appointed conchologist and palaeontologist at the Auckland War Memorial Museum.

When I was a small boy, my family generally spent the summer months at Takapuna in a house on Hurstmere Road with a beach frontage. We bathed and learned to swim, dug and built sand-castles, ate pipi, fished for piper and pakiti, and explored 'round the rocks' of basalt towards Milford, less often under the Waitemata sandstone cliffs towards Cheltenham where we knew the residual stacks on the rock platform as the King and Queen, the names used by Bruce Mason in his *The End of the Golden Weather*. The habit of bringing fish, crabs, shrimps and shellfish home in a tin bucket of sea water began early (say 1921), and within a year or two I was encouraged to start a shell collection, using some small wooden cabinets of drawers dating from Victorian times, and helped by my mother printing labels in a neat hand. (Shells did not smell as badly as dead fish or crabs!)

My father's books included a copy of a popular edition of Cuvier's *Animal Kingdom* ... with pictures of *Conus* in colour and other shells as engravings. At this date, Moss's *The Beautiful Shells of New Zealand* (1908) was out of print, and (short of buying Suter) we had no real guide to the names of New Zealand shells until Dr C.R. Bucknill's book *Sea Shells of New Zealand* was published in 1924, with illustrations mostly by A.W.B. Powell. I received a copy of Bucknill for my ninth birthday … and labelling proceeded apace. I can see in my mind's eye my mother's neat lettering '*Barnea similis*' in blue fountain pen ink on a plain visiting card, neatly underlined in red ink. We made occasional visits to the old Auckland Museum in Princes Street, but found this of little help. Nor, indeed, do I recall much encouragement from school until a later date.

Some forgotten contact of my parents put me in touch with Mr Alfred Suter, son of the writer of the *Manual of the New Zealand Mollusca* … Alfred Suter, by then a middle-aged man, had collected land shells for his father and enjoyed taking me out as a companion to Titirangi or the Waitakere Ranges, to search for tiny snails in the forest litter. Our dentist was Dr Holbrook A. Chatfield, a sportsman and naturalist who took a deep interest in the Auckland Institute. He gave me the duplicates in his collection of *Transactions of the New Zealand Institute*, early volumes with romantic reading of Colenso, Travers, Hutton and others, who whetted my taste for many different topics. I recall he gave me a *Calliostoma selectum* and an *Alcithoe depressa* from Ninety Mile Beach, rare species for an Aucklander who could not wander far from base.

In one of the school holidays … my parents drove me to Tauranga for a few days, partly because Dr Bucknill's book recorded so many species from 'Mount Maunganui' and partly to meet the old man himself. This must have been only a year or two before his death. Naturally we could not collect a fraction of the species he had obtained in many years' residence at Mount Maunganui, but the experience was memorable and Dr Bucknill urged me to get to know Baden Powell. I was too shy, but I already knew the antique shop Mr Powell's parents kept at the top of Shortland Street, and I occasionally bought a shell there when I had been given money as a present. Bucknill also told me I should supplement his book by using Suter's *Manual*, which I was given for Christmas, 1927.

My parents were friendly with Charles R. Laws … [who] returned from war service to study for his degree part-time at Auckland University College, while training and working as a teacher, living at the bottom of Minnehaha Avenue, Takapuna. He had become interested in shells from his geology classes and collected both living and fossil molluscs. When his wife had twins, my mother gave Mrs Laws a chance to recuperate by looking after the babies for some weeks, and on our next stay at Takapuna I was invited to call and see the Laws' shell collection. This neatly curated collection had a tremendous influence in inspiring me to a more scientific approach to my collection and field work, so that I read everything I could get on mollusca and began to preserve Nudibranches (etc.) in methylated spirits (then available without coloured dye). I think it was at this stage … that my father (who had been a member of the Auckland Institute for years) asked for his entitlement of the annual *Transactions* volumes, so that I was able to read (but not necessarily to digest) the papers by Finlay, Marwick, Bucknill's Chiton papers and a paper by Farnie (1919) on the anatomy of *Amphibola crenata* that had led me to dissect not only that species but also the large *Haminoea* we used to collect at Waikowhai on Manukau Harbour. The contact with Charles Laws (who became Dr C.R. Laws, a lecturer in geology at Auckland University) also made me familiar with fossil shells.

I used to read the nature columns that appeared in the Saturday supplements to the *New Zealand Herald* and *Auckland Star*, and for some time I corresponded with Mr A.T. Pycroft,

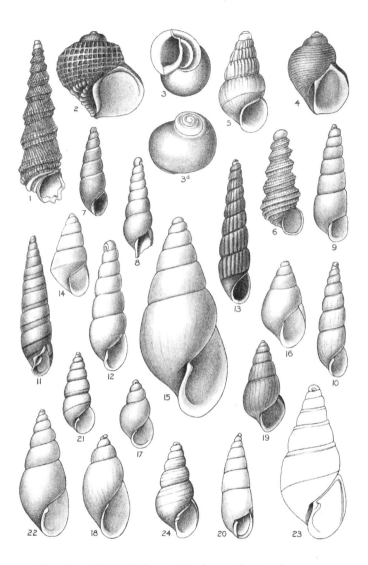

Henry Suter (1841–1918) was a Swiss-born zoologist and palaeontologist whose *Manual of the New Zealand Mollusca* (1913), and its accompanying *Atlas of Plates* (1915), described more than one thousand species and was the standard reference text used by Charles Fleming and generations of other shell collectors. This is a reproduction of Plate 16, a selection of marine gastropods.

who wrote for the latter paper, but I signed my name as 'Riroriro'. His son Leigh went to school with me and I met 'Pyc' when I visited his home, a link with someone who had corresponded with Sir Walter Buller and who seemed to know everyone who was active in natural history studies. By this time (1929) the Auckland War Memorial Museum had opened and I spent many Sunday afternoons comparing specimens from my collection with Powell's new display cabinets.

I began secondary schooling at King's College in 1930, where I was lucky enough to have the late Mr William Delph as a teacher in my first term. At that time he was working with Gilbert Archey on a description of Piranui Pa, near Arapuni. He picked (or had been told) my interest in natural history. When Mr Court offered an excursion on his launch *Ruamana* on March 15, 1930, 'Bill' Delph … asked me … to join more senior pupils on a Field Club trip to the fossil bed near Oneroa, Waiheke, on which Mr Powell and Professor J.A. Bartrum had published the first paper a year before. Powell and Pycroft came along as guests and leaders. After we came ashore, most of the party walked off to the fossil bed, and probably I would have done likewise if I had known where they were going. But I stayed at the water's edge collecting shells in a salt bag and a Mason jar I'd brought with me.

When the party returned, Mr Delph asked me what I'd been doing and (when I told him) he asked me to show what I'd got. I told him I'd got a black Nudibranch I hadn't seen before, 'but I'll find its name when I get home, in Suter's *Manual*'. Immediately a big man in shorts, standing in the shallow water near the dinghy, spun round on his heel to see this odd child who used Suter's *Manual*. When he saw the black Dorid, however, he had to admit that he too had never seen one like it. So he spent the next ten minutes persuading me that it would be best if I let him take it for the museum collections and make a watercolour sketch before it died. It proved to be the second New Zealand record of *Dendrodoris nigra* (as it is now called) but, far more important, it led me to fifty years of friendship and inspiration from Baden Powell.

Later in the same year, Mr Powell … founded the Auckland Museum Conchology Club, now the Conchology Section of the Auckland Institute, which put several young Auckland shell collectors in touch with each other and gave us the chance to learn a great deal more about the mollusca than we could ever do on our own. In addition to weekly meetings at the museum on Wednesday afternoons after school, we had periodic weekend field excursions, among which I remember most vividly those to Rangitoto and Tiritiri Matangi Islands and to Whatipu, Manukau Heads.

From 'Boyhood Recollections of Shell Collecting in the Nineteen Twenties' by Charles Fleming. From *Poirieria*: Conchology Section, Auckland Institute and Museum, 1981.

The Two Lucies

Lucy Moore (1906–87) and Lucy Cranwell (1907–2000) were friends and colleagues who made significant contributions to New Zealand botany, palynology (the study and analysis of fossil pollen grains) and ecology. Having met at Epsom Girls' Grammar School, each graduated from Auckland University College in 1928 with a master's degree in botany. Moore later worked for the botany division of the Department of Scientific and Industrial Research and wrote several important books on New Zealand plants. Cranwell moved to the United States, where she became an expert in palynology at the University of Arizona.

In a 1985 lecture to the Auckland Botanical Society, Lucy Moore looked back at the work she and Cranwell did in the 1920s.

For botany in New Zealand, the period 1925 to 1928 was an exciting time. The second edition of Cheeseman's *Manual of the New Zealand Flora* came out in 1925, two years after the death of the author. We were lucky to come fresh to the new names and concepts. Leonard Cockayne's influence was consolidated by the appearance of the radically revised second edition of his *Vegetation of New Zealand* in 1928, following on the third edition of his *New Zealand Plants and their Story* in 1927. R.M. Laing's *A Reference List of New Zealand Marine Algae* updated this group of plants from Hooker's *Handbook* of 1867. In 1925 the University School of Forestry opened in Auckland, only to close again in 1930. As no biology was taught in boys' schools it was a novelty for us to see young men taking a professional interest in plants. DSIR was set up in 1926, opening up some prospects for employment of future graduates. Donald Petrie, an outstanding contributor to New Zealand botany since 1879, lived his last years in Epsom but we were unlucky not to meet him before he died in 1925. Harry Carse, another notable amateur, made himself an authority on grasses and sedges and we used to tram out to Onehunga to pick his brains. His comprehensive herbarium, now incorporated in CHR,[*] seemed to occupy half of his very small retirement home.

Cockayne's vigorous correspondence with famous botanists all over the world facilitated publication in overseas journals, where a surprising number of New Zealand papers appeared. Also, he was instrumental in bringing exotic visitors here. J. P. Lotsy, the Dutch botanist, lectured on the place of hybrids in evolution. The Swedish couple Einar and Greta Du Rietz stayed six months in the summer of 1926–27, collecting from the Far North to the subantarctic islands, and paying special attention to the lichens. The director of Kew Gardens, Sir Arthur Hill, came in 1928. Such celebrities were considered to be beyond the horizon of students and we missed them all. One we did see was Arthur Sledge from Leeds, who has never forgotten how he ruined his good English boots and lost his pipe on Rangitoto. All the same he retained a great fondness for New Zealand which he handed on to his students of whom one, John Lovis, is now professor of botany at Canterbury.

Perhaps the first two-day trek that Lucy C. and I made together was to Tamahunga (436 metres), the hill at the back of Leigh. We knew nothing about tramping or tramping clubs, and certainly had no backpacks. We carried our bedrolls across our shoulders— very awkward for pushing through bush! After a night at the trig we stumbled down the track carrying a shelf fungus 40 centimetres or more across, trying to preserve its pristine whiteness. I wonder if it is still in the museum.

---

[*] The CHR Herbarium in Lincoln, Canterbury, founded in 1928, and now known as the Allan Herbarium.

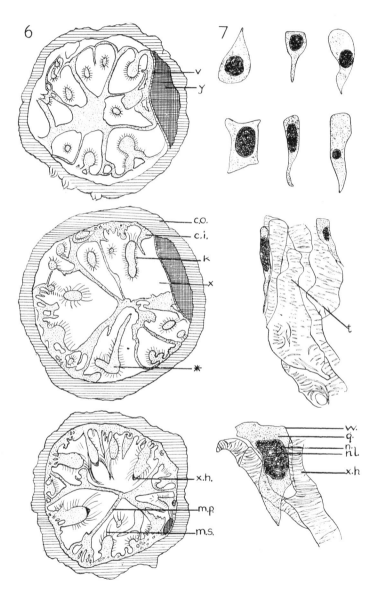

Lucy Moore's cross-sections and illustrations of a rhizome of the root parasite *Dactylanthus taylori*. *New Zealand Journal of Science and Technology*, 1940

Another very early field excursion was to the King Country in search of *Dactylanthus*, the subject of my thesis. Sergeant Fearnley of the Te Kuiti police was our contact and we travelled some way under his escort on the back platform of a railway carriage, looking as disreputable as any pair of prisoners but eagerly learning from him about local plants. From Mangapehi we two went eastwards, riding on a horse-drawn trolley that ran on wooden rails, to the Ellis & Burnand mill and into untouched parts of that marvellous podocarp forest—a never-to-be-forgotten sight, the like of which will probably not be seen ever again.

In 1929 we were graduates without jobs but we were lucky to be awarded jointly the Duffus Lubecki Scholarship … and with its forty pounds or so we undertook to investigate the vegetation of Moehau at the tip of the Coromandel Peninsula. The unexpected assemblage of alpine plants on the summit had been recorded in 1888 by James Adams … but there were no later reports; the list of herbfield plants in Cockayne's *Vegetation* was quoted from Adams. At Easter 1929 we set out to reconnoitre the mountain. Just to get there was an undertaking: NSS Co. steamer to Coromandel, service car with Dick Goudie to Colville and even a few miles further by the clay road that would be impassable if wet. From the hill within sight of Port Charles it was a matter of walking, carrying everything we had, as far as Stony Bay. The leading ridge from there gave the most direct access to the Moehau trig (892 metres), where we had our first glimpse of mountain plants; guided by one of the Bronlund boys we descended by another route. Then we pushed on by the coastal track to Fletcher's Bay, where Captain Fletcher's account of Maori canoe relics in a nearby swamp diverted our attention from botany towards archaeology. The carved prow and stern-piece that were exhumed by LMC[*] are in safe keeping in the Auckland Museum.

---

[*] Lucy May Cranwell.

It was in 1929 too that Lucy Cranwell was appointed botanist at the museum, which was just settling into its new building, opened November 1929, in the Domain. Here she faced tremendous tasks. The Cheeseman herbarium badly needed attention after being stored in the basement of the town hall, if I remember correctly. Dozens of glass cases yawned empty awaiting exhibits. Public relations were very important as the museum depended largely on voluntary contributions from local bodies and private donors for funds and materials. The staff was very small and the half-dozen professionals were expected to carry out fieldwork and to produce papers for the burgeoning museum records. A big series of challenges! But she didn't give up the Moehau project. …

Dr Cockayne was eager for information and full of advice. I can quote from a letter he sent to LMC on 7 May 1929. We must take an aneroid, a film camera and stand; he wants a vivid description and general photos of summit vegetation and close-ups of the margin of the forest, and to know about altitudinal belts; a list of all species observed from base to summit and some idea of relative abundance; on the summit to find out the nature of the soil; relation of plants to wind, shade etc.; how are the species combined together; the density or openness of vegetation; life-form of each species, in detail; relative abundance of seedlings; general colour; does forest become stunted? are there any rocks? We were to collect living plants and bring pieces of all the summit species 'for growing in my garden so that they can be compared with other forms'; bring plenty of dried specimens; for unknown or critical species take only one piece from each of a number of different plants; pay special attention to suspected hybrids; don't neglect grasses and sedges; be sure to collect all forms of *Alseuosmia*. Don't rely on memory—have notebook in hand all the time; in the evening write up a general account of the day's work from memory and field notes (not easy as we carried no tent). The very next day another letter came, wanting cuttings from sea coast to the top; hybrid seedlings for growing; dried specimens of *Senecio myrianthos* (a species we never found); look out for hebes; are there two species of *Quintinia*? What is the *Dracophyllum*? and *Corokia buddleioides*, what is it like? This made a long list of desiderata but after some 40 field-days over five years we knew answers to most of the questions, though some still remain in doubt.

That year we went three times to Moehau and, as we'd never been on any other mountain, we went one weekend in September to Te Aroha, where C.E. Christensen, a protégé of Cockayne's, saw us well started up the mountain track. Near the summit some white stuff amongst fern patches puzzled me. It was in fact the frozen remains of a snowfall and the night we spent at the trig station was memorably cold—our unlined canvas sleeping-bags were white with frost by morning. It was in this quite full year too that we began a long series of tramps out to Anawhata, where the 'university shack' was being built by the half-dozen student owners, of whom LMC was one, and their impecunious friends. This small building and the beautiful headland property were later donated to the university.

From 'Auckland Botany in the Cranwell Era' by Lucy Moore, *Auckland Botanical Society Newsletter*, July 1986.

Tutira: The Story of a New Zealand Sheep Station

Herbert Guthrie-Smith (1862–1940) was a Hawke's Bay sheep farmer and naturalist, and is best known for his classic environmental history *Tutira: The Story of a New Zealand Sheep Station*. This enduring book, first published in 1921, chronicles the natural history and environmental changes on a piece of Hawke's Bay land that Guthrie-Smith spent more than five decades farming. Guthrie-Smith was an astute observer of the introduced weed and pest species that were encroaching on farm and bush land alike. In this passage, he chronicles the spread of the blackberry, a weed he defines as a 'pedestrian' species that travels 'by repeated portages over short distances, re-establishments again and again for another and another step inland, up-country, Tutirawards'.

From northern ports, too, such as the Bay of Islands and Gisborne, pedestrian weeds have also reached the station; accommodation paddocks of roadside inns, drovers' camps, and Maori villages have, as in the south, proved their chief recruiting-grounds and multiplication centres.

Of these pedestrians, the blackberry (*Rubus fruticosus*) was, if not the earliest, one of the earliest to move inland. It stands forth —that fatal and perfidious plant, sown in the eclipse and dug with curses dark—as the single alien that is the master of the sheep, the one plant that makes a victim of him. Its normal habit in the open is to grow into an oval bush. Specimens thus shaped expend their energies harmlessly or comparatively so, although each season the base of the bush increases. Should, however, one of these tall cones be burnt, spread is accelerated laterally; huge horizontal shoots are sent forth, tentacles by which the victim is seized. A sheep but newly caught and still but loosely gripped exhibits an instance of inert brainlessness almost unimaginable; although one determined pull would free the animal, he yet suffers himself to remain anchored by a single strand. Tethered thus, further entanglement is but a matter of time; wool and bramble shoots become woven and twisted into a rope, until finally the sheep dies and its carcase goes to feed the triumphant plant. Perhaps unlimited time only is required to develop out of *Rubus fruticosus* a sheep-catching plant with more enormous shoots and yet stronger thorns.

No good word can be urged for the unhappy plant; not even its fruit, borne in vast quantities but lacking flavour, can excuse or even condone its iniquities. How and when the blackberry reached New Zealand I know not. Its importation is often, I believe, erroneously ascribed to the much-abused missionary; certain it is that the weed has not come into Hawke's Bay from the north. Its local origins are Petane, and Tangoio, where long prior to my time stretches of blackberry hedge had been planted.

We can now follow inland the march of this terrible pedestrian. After leaving Petane the road for some distance ran parallel to one of these planted fences, a brazen example of a vested interest, for when at a later time blackberries were attacked with poison and spade, this hedge, grey in its hoary iniquity, was spared. There were several bushes scattered about the sandy hummocks of flood-silt in the Esk river bed. Throughout the native cultivations, where there are now hundreds, I do not remember a specimen; two there were, however, on the shingle flats near the Coastal Hill. On it were established other two bushes: a single specimen grew at the base of Pane-Paoa, the County Boundary Hill; another huge plant grew where the road strikes sharply inland from the beach. Between that and the Tangoio homestead another hedge had been deliberately planted, seedlings dibbled in at regular intervals. Blackberry bushes were scattered here and there about the Tangoio homestead and along the bridle track till it began to rise to the hills. Halfway between the Tangoio Flats and First Fence there was one bush. Between First Fence and Kaiwaka boundary gate there were none, and but a single stunted specimen on the high pumiceous tops. On the limestone edge overlooking the Waikoau valley flourished the furthermost inland centre of mischief, a colony of six or seven immense bushes. Another blackberry grew within half a mile of the crossing, another immediately on the Tutira side of the ford. There were none on the site of the disused Maori cultivation grounds on the Racecourse Flat, pretty good evidence that the plant was a genuine pedestrian sticking to the road, that it had not been deliberately brought up as a fruit, and finally, that it could not have been in the province at an early date. There was a plant on the old native trail halfway to the Maheawha crossing, another at the ford itself. The westernmost bush on Tutira proper was established just above the gorge separating Tutira and Putorino.

During the early 'eighties, in fact, except about the plague-spots Petane and Tangoio, blackberries could almost be reckoned on a man's fingers.

There were, however, even at that date, dotted along the road, bridle track and pack trail, a sufficiency of bushes to fix it definitely as a line of human traffic. The pioneers of the east coast had in fact marked their pilgrim path in blackberries, for it is man himself who first carried up country the fatal seed. Each offering deposited at each improvised temple of Cloacina on the road has erected itself a living monument to the goddess; whilst intermediate bushes could still be individualised, they were to be found more thickly in proximity to the parent plantations, more sparsely at longer, or as I may say, more costive distances. Owing to its ensnarement of sheep, the blackberry is the most dangerous, perhaps the one truly dangerous, alien in New Zealand. On hill land impossible to plough on account of gorges and landslips, the only method of eradication is by spade and mattock; even then these diggings have to be gone over again and again: the smallest rootlet grows, even half-buried leaves will root strongly in damp spots. The plant, moreover, possesses an intelligence and energy worthy of a better cause. Again and again I have dug out bushes, especially on light lands, sending forth roots which a few inches beneath the surface have followed exactly the lines of sheep tracks within range—tracks enriched by manure carried from contiguous camps. Removal of the soil has revealed a subterranean root system corresponding to their sinuosities. It only remains to add that after sheep had acquired a taste for the fruit—I have seen their paunches black with the berries—and more especially after the arrival and increase of imported birds, who carried the seed everywhere, the bramble increased in a most alarming way. Although a fortune awaits the inventor, no weed-destroyer has yet proved efficacious. It is impossible not to look with grave concern at the future of many hundred thousand acres in northern Hawke's Bay.

The blackberry continues to be a widespread pest in Hawke's Bay. Guthrie-Smith was also concerned with the exotic insects that made their home at Tutira.

In the open the mason fly plasters its cells on to the pitted surfaces of limestone crag; within doors its vermiculated clay chambers are fitted into every available crack and chink, into keyholes, beneath projecting laps of weatherboarding, in folds of suspended garments. A situation particularly favoured is an oilskin coat suspended on a verandah—such an article, if shaken after prolonged disuse, always precipitating a rain of broken clay chips and flaccid spiders. Every chamber contains cells of different sizes, in each of which an egg is deposited, and the compartment then filled in with spiders, which for long retain their freshness, and which appear to be torpid rather than dead. In due course the eggs hatch, and the grubs feeding on the stores provided become white maggots. Later again—unless, as not infrequently happens, destroyed by parasites —the mature insect, dark, slender, and elegant, emerges and completes the circle of life.

The black cricket, *puharanga*—'bushranger'—of the natives, whose faint musical trill tells us that autumn has come once more, is reputed to have reached New Zealand either in matting from the islands, or in the bedding of troops from India. It has never been plentiful on Tutira; the rainfall is too great for a semi-tropical insect, the soils of the run too porous. Only in localities where alluvial clays fissure and crack in summer can the insect become a plague, but on such lands I have known its numbers multiply into millions.

The honeybee was liberated about the same date at the Bay of Islands and at New Plymouth; probably from the former, Hawke's Bay has been stocked directly or indirectly. The newly imported insect had no enemies to contend with; there were no diseases and no competitors. The winters of the North Island are brief, or non-existent; there is no single month of the year when some native shrub or another is not to be found in blossom. Local conditions were extraordinarily favourable too; portions of eastern Tutira, viewed even from considerable distances, were during springtime actually grey with the profusion of white clover heads. Everywhere then also the purple-headed prickly thistle possessed the land. There was not a hollow tree or crannied limestone rock which in the 'eighties did not contain a hive; colonies were established even in the open, though from these unsheltered swarms no great store of honey was obtainable, dews and rains diluting the nectar gathered, and washing it from the uncapped cells.

The exuberant prosperity of the bee has passed away with the disappearance of the white clover and the thistle. Few indeed of the hollow crags now harbour colonies. One rock only—a vast square projecting from the highest tier of ocean floor on the Racecourse Paddock—has never to my knowledge during forty seasons been untenanted. Bees are now again on the increase, owing to ploughing, the use of artificial manures, and consequent revival of white clover.

> Blackberries, mason flies, black crickets and honeybees were already at Tutira when Guthrie-Smith took over the farm in 1882. While farming the run he was witness to the arrival of many other species. When rabbits and weasels arrived in Hawke's Bay, he said 'better that a millstone had been hanged about that district's neck and that it had been cast into the depths of the sea.' The weasels, part of the 'invasion from the south', arrived in the district at the beginning of the twentieth century.

Prior to their arrival at Tutira, during their approach through the southern settled districts of Hawke's Bay, terrible tales of the murder of young lambs, of the biting of babies and grown folk, of rape of hen-roosts, were rife in the daily papers, perhaps—for squatters had imported the horrid vermin—most prominently in papers hostile to the sheep-farmer interest. At the time I took considerable trouble in the investigation of several of these stories of attacks on grown folk, and believe that some at least were true, or at any rate that much evidence of a circumstantial sort could be adduced in support of them.

The earliest weasel was seen on Tutira in 1902. Between that date and 1904 they had overrun the country between Tutira and the southern edge of the Poverty Bay Flat. Everywhere I heard of them. On every road and new-cut bridle track during these two seasons I met or overtook weasels hurrying northwards, travelling as if life and death were in the matter. Three or four times also I came on weasels dead on the tracks. These weasels, alive or dead, were or had been travelling singly. The only party I heard of was reported by Mr J.B. Kells, then managing Tangoio. In firing a small dried-up marsh he dislodged a large number; according to his statement, they 'poured out' of the herbage. For a short period weasels overran like fire the east coast between Tutira and Poverty Bay, and then like fire died out. I traced them by personal observation to the very edge of the Poverty Bay Flats, then, like the Great Twin Brethren, 'away they passed and no man saw them more'. Nowadays on Tutira I do not hear from shepherds or fencers of the weasel once in six years. I have not seen one for twenty years. There is something ridiculous in the fact that the weasel should have arrived on the station before the rabbit, and that later, when rabbits had become numerous, weasels should have practically passed out of the district—that the cure, in fact, should have preceded the disease.

From *Tutira: The Story of a New Zealand Sheep Station* by H. Guthrie-Smith, second edition: William Blackwood & Sons, Edinburgh and London, 1926.

Beginnings: Guthrie-Smith in New Zealand, 1885

Herbert Guthrie-Smith bought the lease to the sheep station at Tutira in 1882. The land was marginal and the farm unprofitable, but after years of hard work clearing land and planting grass seed, Guthrie-Smith turned the farm around. In this poem, written nearly a century later, Peter Bland evoked his struggle.

Who am I? What am I doing here
alone with 3000 sheep? I'm
turning their bones into grass. Later
I'll turn grass back into sheep.
I buy only the old and the lame.
They eat anything—bush, bracken, gorse.
Dead, they melt into one green fleece.

Who am I? I know the Lord's my shepherd
as I am theirs—but this
is the 19th century; Darwin
is God's First Mate. I must keep
my own log, full of facts if not love.
I own 10,000 acres and one dark lake.
On the seventh day those jaws don't stop.

Who am I? I am the one sheep
that must not get lost. So
I name names—rocks, flowers, fish:
knowing this place I learn to know myself.
I survive. The land becomes
my meat and tallow. I light my own lamps.
I hold back the dark with the blood of my lambs.

From *Stone Tents* by Peter Bland: London Magazine Editions, London, 1981.

The Maori Way of Birth

Peter Buck/Te Rangi Hiroa (*c.*1877–1951) studied medicine
at the University of Otago, and went on to have a career as
a medical doctor and member of parliament for Northern
Maori. During the First World War, he initially served as a
medical officer but rose to become second in command of
the New Zealand Pioneer Battalion.

After returning to New Zealand as director of Maori
hygiene for the Department of Health, Buck turned increasingly to anthropology, eventually becoming a leading
anthropologist of Maori and Pacific peoples at the Bishop
Museum in Hawai'i. In *Vikings of the Sunrise*, first published in the United States in 1938, he wrote a lively and
accessible account of the culture and traditions of Pacific
peoples. In the chapter on New Zealand, he told his own
story.

My mother was a full Maori of the Ngati Mutunga tribe of North
Taranaki in New Zealand. She had the arresting name of
Ngarongo-ki-tua (Tidings-that-reach-afar). I hope, for the sake
of her memory, that by gathering tidings from afar I may be
worthy of the honour of being her son. She was the firstborn of
the senior family of the Ngati Aurutu sub-tribe, and I absorbed
pride of race from her. Her only brother was named Te Rangi
Hiroa after an ancestor who had lived two centuries before. I was
told that Hiroa was a contraction of Ihi-roa and that the name
meant the Heavens-streaked-with-the-long-rays-of-the-sun. My
uncle became seriously ill during a visit to a distant village and
commanded that he be moved in order that he might die at
home. Unfortunately he died on the way, and I was given my first
name of Te Mate-rori (Death-on-the-road), a wretched name
because 'rori' is the modern Maori form of road. I was greatly

relieved on reaching my teens to be given my adult name of Te Rangi Hiroa in more classical memory of my uncle.

My father belonged to a north of Ireland family that lived in Armagh, so I am entitled to his family name. I am binomial, bilingual, and inherit a mixture of two bloods that I would not change for a total of either. I mention this brief family history to show that from my birth I was endowed with a background for the study of Polynesian manners and customs that no university could have given me. My mother's blood enables me to appreciate a culture to which I belong, and my father's speech helps me to interpret it, inadequate though the rendering be at times.

My maternal grandmother was a wonderful old lady. She had lived so long that she had acquired more wrinkles than anyone I have ever seen. She had seen many of our tribe die, and she had mourned over them all. It used to be the custom when wailing over near of kin to incise the skin with a flake of obsidian so that the flow of blood and tears might mingle to the fullest expression of grief. Sometimes charcoal was rubbed into the cuts and left indelible marks. My grandmother's breast was covered with such grief marks; and for her very dear ones, she had made the record on her cheeks. I was particularly proud of her tattooing. She had the orthodox pattern for women on both lips and her chin was covered with an artistic curved design. But in addition she had beautifully executed double spirals on either nostril and short curved lines on her forehead that arched upward from the inner angles of her eyes. When I was chastised at home for some error in conduct, I ran away to the Maori village and took refuge with my grandmother. She told me tales of happenings in her girlhood, and I learned tribal history from her as well as from my mother.

When I went to Te Aute College after my mother's death, I spent some of my vacations with my tribe. In spite of the protestations of my relatives with houses of sawn timber, I insisted on sleeping beside my grandmother on the earthen floor of her native hut with its walls of tree-fern slabs. She grew her own tobacco outside her hut, and, as she smoked her pipe beside her charcoal fire in the evenings, she told her college grandson stories that it was a privilege to hear. She had been an eyewitness of so much that had passed that she belonged to another age. With each parting, she wept the longer, and we both realised that the end of our companionship was approaching nearer and nearer. She has passed away to join her daughter in the Polynesian spirit land, and I would that our myths of that land were true. Her name was Kapua-kore which means Cloudless, an apt name for one who in her long life brought no cloud of sorrow to any living soul.

After graduating in medicine in 1904 and spending a year as a hospital intern, I obeyed the call of my blood and joined the government service as a medical officer of health to the Maoris. I visited various villages and was received in all with the courtesy that still takes the form of old-time ceremony. The people gathered in the open space before the village assembly house, and tears were shed for those who had recently passed away. The Maori *tangi* (weeping) and the Irish wake are similar in fundamental principle, and on such occasions my two halves could unite as one. Speeches of welcome couched in archaic form were made by the local chiefs to which I replied as best I could. Five years' study at a medical school with a year in hospital had made a serious break in the continuity of my Maori education. My Maori words unconsciously flowed along an English channel of grammar, and I was horribly conscious that I was talking to my own people like a foreigner. The speeches were followed by the ceremony of pressing noses with all and sundry. This form of greeting, at one time universal throughout Polynesia, now survives as a regular custom only in New Zealand. It says much for our generation that we never tried to evade the custom because we did not wish to give pain to our elders. After these nasal contacts, the taboo of the stranger was lifted and one could mix freely and talk informally.

The visitor was the guest of the village, and the best food was provided for him during his stay in the tribal guest house. Different districts have local foods which are a great asset to the people not only for their own sustenance but for the entertainment of their visitors. Fish, crustaceans, and shellfish in the coastal districts, eels and whitebait in the river regions, pigeons and parrots in the forest areas; all had their particular season when they were at their best. My own district was famous for its lamprey eels in June or July. The sea eggs (echinoderms) were fat at Te Araroa when the golden flowers of the kowhai blossomed in spring. Sharks came into the fishing grounds off the Taranaki coast when the new growth of bracken fern began to straighten out its curled shoots. I learned to know the food seasons of the various parts of the island, and I tried to make my visits of inspection coincide with the native food calendar, not only because I liked native foods but because native hosts were so genuinely pleased to lay before their guests the foods for which their district was noted. Economic embarrassment was avoided, and host and guest shared a common satisfaction.

I early realised that to gain the interest and support of chiefs and leaders older than myself, I must overcome the handicap of youth by an exhibition of Maori scholarship that would not only earn their respect but indicate clearly where my sympathies lay. I commenced an intensive study of Maori mythology, legends, traditions, and the details of customs, manners, and etiquette. I learned the pattern of ceremonial speech and the forms of metaphor and simile that went with it. The more a speech is illustrated with quotations from myths and ancient traditions, the better a Maori audience likes it. Old songs and incantations with an apt bearing on the subject matter are necessary because a speech is regarded as incomplete without them. I was never good at rendering songs, but I acquired a host of chants and incantations to illustrate speeches. I combed the printed literature, and I learned at firsthand from the experts of various tribes who were only too pleased to impart their knowledge to an appreciative student of their own blood. With others of the younger leaders, I became a home-made anthropologist—not to obtain a university degree, but to gain an inner understanding of our own people in order that we might the better help them through the problems and trials created by civilisation.

> Buck did become an anthropologist, and eventually became more well-known for that than for his medical work. This passage entitled 'Birth and Infancy' is from his 1949 book *The Coming of the Maori.*

The children of commoners were born with little trouble or fuss, except to those immediately concerned. They had no inheritance of *mana* and tapu and no prospects of power and prestige. As the tapu of blood was equally obnoxious to poor and rich, confinement took place in the open or in a rough shelter instead of in the dwelling-house. The father and mother of the patient were in attendance with the husband. During labour pains, the Polynesian squatting position was assumed and the patient supported herself by holding to hand posts erected for the purpose. The mother or some other person assisted by clasping her arms around the patient's abdomen from behind and helping the bearing-down pains with external pressure. Usually there was no difficulty, as the squatting position was the best for directing the contractions of the womb in the right direction. The proximal end of the navel cord was gently massaged towards the abdomen, and the cord tied with flax fibre. The massaging was supposed to prevent a protruding navel. The placenta was buried somewhere without any ceremony, and the dried cord, when it separated from the navel, was also concealed.

Childbirth among the aristocracy was a different proposition, and interest spread among the people when it was known that the chief's wife was pregnant. The vagaries of appetite which occur during pregnancy have led to consequences of historical importance. An example was furnished in the story of the *Aotea* canoe in which the pregnant Taneroroa induced her husband to steal her elder brother's dog to provide the dog's flesh for which she hungered. The incident led to enmity between the two families and the removal of Taneroroa's family to the territory north of the Patea River. The Ngarauru descendants of the elder brother, Turangaimua, state that when the Ngati Ruanui descendants of Taneroroa become excited, they use the word 'au' as an exclamation. Then with a smile, the Ngarauru say, 'That is the bark of the dog which their ancestress stole.'

When the expectant mother felt that the confinement was drawing near, a maternity house termed a *whare kohanga* (nest house) was built in some secluded spot away from the village. Female attendants termed *tapuhi* waited upon her, and other servants attended to the cooking in the open or in a roughly constructed shelter. The maternity house was tapu to all but the personal attendants, hence the cooks brought the food part way and the attendants took it into the house.

In some families, the services of a priest were necessary to recite appropriate chants to Hineteiwaiwa, the goddess of parturition. Certain chants were recited over a normal delivery, others over delayed labour to promote the opening of the path by which the child would emerge into the world of light. Best … gives a slightly different posture and treatment, in that the patient knelt down with the knees wide apart, while the female attendant squatted before her. They clasped one another under the armpits, and the attendant used her knees to make pressure against the abdomen from above downwards to assist expulsion.

The cutting of the cord (*iho*) was treated as a minor operation without any special ceremony such as that held in the Cook or Society Islands. Lacking the bamboo knife of Polynesia, the instrument used was a sharp flake of stone or chip of obsidian. The operator was one of the female attendants or an experienced female relative included among the attendants. After gently massaging the attached end of the cord, it was tied close to the abdominal wall with flax fibre or some other material such as the thin stem of a creeping plant (*makahakaha*) which had been scraped and soaked in water to render it pliable. The cord was cut about an inch and a half from the navel, or the measure from the tip of the thumb to the nearest joint (*konui*), or sometimes the longer measure of the length of the little finger (*koiti*). The cut end was smeared with oil expressed from the seeds of the titoki (*Alectryon excelsum*), and a dressing of the inner bark of the lacewood (*Hoheria sexstylosa*) soaked in titoki oil was applied. The dressing was kept in place with a bandage formed of a wide strip of inner lacewood bark, which readily splits into thin ribbon-like layers. During the daily washings, the cord stump was examined. Any necessary swabbing was done with flax tow (*hungahunga*) obtained by scrapings of flax fibre (*whitau, muka*). The afterbirth (*whenua*) was buried where it was not likely to be walked over. The dried cord, after it separated from the navel (*pito*), was placed in the cleft of a rock or tree or buried on a boundary line or elsewhere, and the spot was marked by a stone or post. The place was referred to as the navel cord of whatever the name of the child (Te Iho o . . .). If a tree were used, the whole tree received the name, the well-known hinau tree named Te Iho o Kataka, for instance. The method of disposing of the cord in the cleft of a rock or a cliff was common throughout Polynesia.

One of the first things the mother did after giving birth was to bathe in the stream or river to cleanse herself physically and psychologically. She was then ready to receive a ceremonial visit from representatives of her family and her people to welcome the new addition to the tribe and to congratulate the parents and their family. The ceremony of welcome was termed *koroingo* or *maioha*, and it was particularly observed on the birth of a first-born who had been announced by the sounding of the shell trumpet. The mother, with her child in her arms, sat in the doorway of the maternity house and the relatives of the father and the mother deposited their gifts separately in the open space before the house. The leaders of either side made speeches of welcome to the relatives of the other side and then to the child. The greeting to the infant was augmented with old-time chants or songs which referred to the mythological creation of man, the development of the child within the mother's womb, and the final emergence into the land of the living. The speeches concluded the formal visit to do honour to the newly arrived member of the tribe.

After the social visit, the mother resumed her normal activities in a much shorter period than the ten days' rest insisted upon by European physicians for their patients. When the mother and child returned to the village, the maternity house, which was tapu, had to be destroyed to prevent it becoming a source of danger to others. The woodwork, if used as firewood for cooking fires, would result in affliction and death. Later, even any material which had grown on the site would bring trouble to those who used it. Thus as a preventive measure, the house was burned down, as were any mats or material used within it. A priest conducted the ceremony to remove the tapu from the house site. From a modern point of view, the maternity house was equivalent to an infectious disease shelter, and the treatment by fire and ritual destroyed a psychological source of infection in a similar way to the material destruction of microbic infection by fire or fumigation with chemicals.

From *Vikings of the Sunrise* by Peter H. Buck (Te Rangi Hiroa): J.B. Lippincott Company, Philadelphia, 1938; and *The Coming of the Maori* by Te Rangi Hiroa (Sir Peter Buck): Whitcombe and Tombs Ltd, Wellington, 1949.

A Man Who Moved
New Zealand

The Alpine Fault, one of the world's longest active earthquake fault lines, bisects the South Island from Blenheim
to Milford Sound, marking the boundary of the Pacific
and Australian tectonic plates. But in 1941, when Harold
Wellman (1909–99) was working as a geophysicist for the
New Zealand Geological Survey, the fault had not yet been
discovered, and the theory of plate tectonics was still two
decades away.

Wellman and his colleague, Dick Willet, spent August
1941 in unmapped territory in South Westland searching
for mica, a strategic mineral used in the manufacture of
radio condensers. Laden with supplies of lima beans, bacon
and macaroni, they travelled by rail to Hokitika, then
hitchhiked south. Wellman tells the story of their discovery
of the Alpine Fault in his typically understated way, not
acknowledging that this was one of the most significant
geological discoveries of the century.

This was when we started to extend the geological mapping into
the unknown. We needed a good view, and the best way to do this
was by getting a lift on the back of an open truck. The front of the
alps is very spectacular, and it didn't take long for us to realise
that the Gregory Valley fault could be readily traced southwards.

Dick had done his master's thesis on glaciation, and I wanted
to show him some of the glacial features I had seen while gold-
mining. We went down to Waiho Beach. Most of the miners had
left, and we had no trouble getting a hut to stay in. The next day
we walked along the coast from Waiho Beach to Gillespies Beach,
where we stayed with Jock Thomson, my old gold-mining mate.
As we looked back to the mountains we took photos, and noted

how the moraine hills that slope towards the coast seemed to be faulted off at the front of the alps. …

The Gregory Valley was even better in South Westland than in the north. It seemed to be an unusually straight major fault separating granite on the north-west from schist of the alps to the south-east. We decided that the Gregory Valley was merely the result of the more rapid erosion of the rocks along the fault, which we later decided to call the Alpine Fault.

There was a large exposure on the track at a place called Blue Slip where we were able to get a good look at the work of the Alpine Fault. The schist was broken in pieces some fifty milli-metres long, but we could see from the schist layering that the individual fragments had moved very little. The broken schist looked like the result of explosions, or perhaps we were seeing the heart of old earthquakes.

We walked down to the Haast River, crossed it on a temporary bridge, and went to the public works camp to get a meal and somewhere to sleep. When the weather was clear we spread out our roll map on the ground, and saw the relation of the fault to the landscape for the first time. To the north we could see the fault along the front of the Mataketake Range, and to the south we could see that it extended along the range front. From the map we could see that it was likely to extend up the Jackson River, over the Martyr Saddle, and then southwards to Lake McKerrow at Hokuri Creek. And if that was correct, it was only another twelve kilometres to Milford Sound and the straight coastline of Fiordland.

A decision had to be made—should we return to Wellington, having completed the search for mica, or should we carry on and see if the Alpine Fault could be traced further south? We decided to carry on, as we told ourselves that Dr Henderson would be pleased if we had the makings of a good map. Once we had obtained extra stores and posted extra gear back to Wellington we headed south. The Jackson River follows along the Alpine Fault, and the position was exactly as predicted. We could also see that the belt of ultramafic rocks was cut off by the Alpine Fault.

The cattle track led down the Cascade Valley, past the Hermitage Swamp to the coast at Barn Bay. We plodded down the coast. The sandy beaches were easy, but the large boulders were slippery. It took us two days to walk along the coast, and we were almost out of stores when we reached Gunn's Hut at the Hollyford. The hut was open but dusty, so we took some stores and left money to pay for them. From there it was only seven kilometres to Hokuri Creek, the place where we had guessed that the Alpine Fault would cross Lake McKerrow. We got there in the early evening, which was fine and clear, and the fault was exactly as predicted—a three-metre scarp. On the opposite side of the lake was a stream right on the line of the fault and a step in the skyline.

We made camp, feeling pleased with ourselves, as we believed that the difficult part of the trip was over. It was not so. It started to rain in the night, and was teeming by morning. Then we discovered that the track just upstream from the Hokuri had been eroded away, and there was a cliff where the track had previously been. We spent half a day trying to get along the edge of the lake, then decided that the track must be higher up so climbed up and finally found it again. Heavy rain continued, and extreme care had to be taken not to be swept away by little streams that normally would be hardly noticed. Two days later, on 26 September, 1940, we arrived bedraggled at the public works camp at Marian for workers on the Homer Tunnel. The following day we caught a bus to Gore, then travelled by rail and ferry back to Wellington.

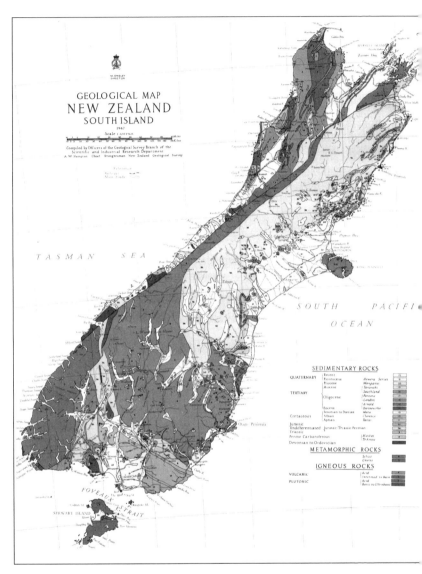

The New Zealand Geological Survey's 1:1 millon scale geological map of the South Island published in 1947 clearly showed the location of the Alpine Fault—and was the inspiration for Wellman's discovery of the fault's 480-kilometre lateral offset.

Wellman and Willett published their findings in 1942, along
with a map showing the position of the Alpine Fault. Seven
years after the discovery of the fault, Wellman made an even
more startling discovery. In 1948 most geologists believed
that all fault movement was vertical, with little or no sideways, or transcurrent, movement. Wellman proposed the
radical idea that the Alpine Fault had moved 300 miles, or
480 kilometres, laterally. As Wellman's biographer Simon
Nathan wrote: 'The idea had come to him on a wet Sunday
afternoon when he was sitting at the dining-room table in
Greymouth, and he immediately took a pair of scissors and
cut a copy of the newly published geological map of the
South Island along the Alpine Fault to see if the opposite
sides matched. There was excellent agreement, and this
seemed to explain all sorts of problems in the distribution
of rocks in the South Island.'

Wellman presented his then controversial idea at a Christchurch meeting of the 1949 Pacific Science Congress. Ross
Taylor, a young graduate student at Canterbury University
College, who was present, later reported to Wellman's
biographer Simon Nathan:

The highlight of the meeting was Harold Wellman. Word had
got around about his ideas, and the room was full. He displayed
a large hand-made geological map (about 3 x 6 feet in my recol-
lection, mounted on an easel) of the South Island, and then,
after talking for a while, suddenly proceeded to slide southern
Westland 300 miles along the Alpine Fault to match up the strata
near Nelson. It was a dramatic moment that I have never forgotten. It made me realise that there was much more to geology
than I had learnt already. Geology at that time was a very con-
servative subject and perhaps the most lasting effect on me was to
hear someone bold enough to put forward such a novel idea.

While most geologists thought it was 'another of Wellman's wild ideas', the 480-kilometre transcurrent offset of the Alpine Fault was eventually accepted. Later, when the theory of plate tectonics emerged in the 1960s, Wellman's brainwave provided concrete evidence that tectonic plates did, indeed, move.

From *Harold Wellman: A Man Who Moved New Zealand* by Simon Nathan: Victoria University Press, Wellington, 2005.

Paradise Lost

Nelson amateur ornithologist Pérrine Moncrieff (1893–1979) was an early leader in the twentieth-century conservation movement, and campaigned tirelessly for the protection of New Zealand's native birds and their habitats. As well as writing popular articles and books—her 1925 work *New Zealand Birds and How to Identify Them* was New Zealand's first ornithological field guide—Moncrieff served as vice-president of the Native Bird Protection Society, and in 1932 became the first female president of the Royal Australasian Ornithologists Union. In her article 'Destruction of an Avian Paradise', first published in Britain in 1944, she documented the decline of native species through the decades of habitat destruction and predation that followed the arrival of Europeans in New Zealand.

It was not long before the outside world discovered that New Zealand was inhabited by birds unique in the history of ornithology. … The world of museums and private collectors decided that all these rare birds were doomed! and deemed it their sacred duty to ensure that specimens of these unique creatures should be secured for the benefit of posterity.

Consequently, the first native birds to become extinct were those which tempted the cupidity of collectors. What a field for Reischek and his contemporaries, who filled their museums with dried and stuffed skins of New Zealand's rarest fauna! Unfortunately, the residents of this country, instead of setting aside sanctuaries in which the native birds might exist unmolested, concurred with the collectors' views and even assisted them to secure specimens. Thus, today, New Zealanders must seek their finest treasures overseas, in the museums of Britain, Austria and America.

The story of the kiwi affords an example of what befell the native birds once their value became known. Long before and after Sir Richard Owen's *Memoirs* [*on the Extinct Wingless Birds of New Zealand*] was published in 1879, an extensive and lucrative export trade in kiwi skins, from both islands, to Britain and Germany was carried on; until, after many years, the British market became glutted. Sir Walter Buller described numerous 'kiwi-hunts', in one of which forty kiwis were captured, in another three hundred. And Lord Walter Rothschild records having seen several sacks of kiwi skins at Stevens' Natural History Salerooms, London. Today the kiwi—ostensibly protected by law—leads a precarious existence, and it is surprising how well it evades its foes. Doubtless the fact that it is nocturnal is in its favour.

Three species of Chatham Island rail were greedily collected, making a considerable reduction in their numbers; and the saddleback (*Creadion carunculatus*), once plentiful, were killed in large numbers to settle a biological argument. Persecuted by collectors and deprived of its forest home, the saddleback became a rarity on the mainland, though still found on certain outlying islands. On one of these rats have become established, and constitute a grave menace to the saddleback; unless it can follow the example of the whitehead—which became almost extinct as a dweller in low trees, but staged a comeback by taking to the branches of lofty trees.

Certain species of birds appear to have been on the decline prior to the advent of the European. Such were the thrush and the kakapo—which, according to the Maoris, fell a victim to wild dogs. Incidentally, the kuri met its fate by 1860 and the vegetarian rat disappeared before the advent of the fierce Norwegian rat.[*]

---

[*] By the 1920s the kiore, or 'vegetarian rat', was widely believed to be extinct. Today the kiore is known to survive in isolated mainland locations and on outlying islands.

Other unique species of birds were no sooner discovered than they made their exit. The Stephens Island wren (*Traversia lyalli*), for instance, succumbed to the lighthouse-keeper's cat; which brought in eleven and then no more. Apparently it had killed out the whole of a species unknown to science.* The discovery of a large rail resulted in the capture of four specimens of the takahe (*Notornis hochstetteri*) between the years 1849 and 1898, after which the brief history of the takahe—whose skin fetched in the vicinity of £250—came to an end. Though it is possible that the world may yet hear of this bird, whose remote haunts have never been thoroughly explored.

Whilst collectors hastened the doom of native birds early settlers proceeded to open up the country. The pressing need was for agricultural land, and a period of rapid clearing of the forest for planting of crops and for pastures ensued. Such clearing was especially thorough in the North Island, until the area of 62,000 square miles, which existed in 1840, has been reduced to some 17,000 square miles at the present day. After sawmillers came pit-prop, firewood and fence-post merchants, who, in early days, ruined the floor of the forest with bullock carts, and in modern times arrive in motor-trucks, attended by cats and pig dogs. The scars thus made in the forest are invaded by bramble, gorse, ragwort and other introduced weeds. In the Urewera country there are some 2,000 acres abandoned to ragwort alone.

Whilst the dictates of civilisation demanded the sacrifice of the New Zealand forest, it is now apparent that, in the light of modern knowledge, many bush areas were unwisely cleared. Yet in 1938, despite all that had been written against cutting down the indigenous forest on high places, sawmillers were operating at

---

* The story of the lighthouse-keeper's cat causing the extinction of an entire species is now considered to be an exaggeration—there were many cats on the island and the lighthouse-keeper's cat can't take credit for the entire job.

an altitude of one to three thousand feet in watersheds and catchment areas which should never have been touched by sawmiller or farmer.

Behind the destruction of the forest lies the disappearance of the native birds of New Zealand. Unable to cope with the change of habitat, those which could not adapt themselves to changed conditions became extinct.

In 1892, Lord Onslow, Governor-General of New Zealand, placed a memorandum before Parliament, directing the attention of his ministers to the fact that under changed conditions of existence many of the native species were passing away; that some had already disappeared, whilst others were verging on extinction. Amongst these was the beautiful huia, of which the chiefs, assembled to commemorate the birth of a son to Her Majesty's representative, said: 'There yonder in the snowclad Ruahine Range is the home of our favourite bird. We ask you, O Governor, to restrain the pakehas from shooting it, that when your son grows up, he may see the beautiful birds which bear his name.' Although the memorandum did much good in that it persuaded Parliament to declare Little Barrier (in the north) and Resolution Island (in the south) bird sanctuaries, unless there exist in some remote corner a few as yet not discovered huia, this bird of curiously restricted habitat, the pride of the Maori race, must swell the list of birds exterminated by man. Fortunately the rare stitchbird (*Notomystis cincta*), which soon gave up its struggle against civilisation on the mainland, still exists on Little Barrier Island. [Whence Andreas Reischek, the collector, removed 150 specimens.]

Had the settlers merely felled the forest it would have been serious, but invariably fires followed the clearing of the bush. … The first victims of fires throughout the country were ground birds, such as the native quail (*Coturnix novaezealandiae*) and the fernbird (*Bowdleria punctata*), of feeble flight. But neither axe nor fire entirely explain the disappearance of the two interesting bats of New Zealand. The long-tailed bat (*Chalinolobus morio*) was quite plentiful in certain localities; but the short-tailed bat (*Mystacops tuberculatus*) was always an extremely rare creature. It has been suggested that they became victims of an epidemic.

It has often been stated that the native birds of New Zealand suddenly decreased wholesale, before the forest was destroyed and vermin had become established. The only explanation of this sudden disappearance seems to be disease introduced from without. It has been suggested that the huia, for instance, was attacked by a tick imported with the Indian myna. Thus it is feasible that native birds were attacked by disease—possibly a form of bird malaria—introduced by European avifauna. Little data is available upon the subject, though it is alleged that traces of malaria have been found in the blood of the native pipit. Like measles amongst primitive tribes, sickness thus introduced may have swept through the ranks of unprotected birds. ...

But for the search for minerals, much back country might have lain unscathed and the native birds had further years of immunity. But the gold fever spread into remote localities and the miners advanced accompanied by their usual companions, dogs, cats and goats—all enemies of the birds. In addition to the above, settlers introduced bees, which, finding the forest more agreeable than their hives, took to the bush and competed with native birds for nectar. An examination of 'bee-trees' reveals only too clearly that the Maoris were correct when they said native birds were being starved by the introduction of bees.

In 1887 [there] appeared a paragraph in the *Southland News* announcing that the acclimatisation of rabbits had proved successful. By the '90s settlers were at their wits' end to know how to deal with these creatures which were threatening whole areas with destruction. In their trouble they imported weasels, stoats, ferrets and polecats: in spite of vigorous opposition from such authorities as Sir Walter Buller, who wrote: 'As shipment after shipment of these vermin arrived in New Zealand I raised my voice in protest against so insane a policy and so did others. … The imported animals were turned loose north and south, and have now become acclimatised in a country where the conditions of life are so favourable … that no power on Earth will ever dislodge them.' … In spite of all that has been said against such vermin, Mr H. Guthrie-Smith placed rats as the worst enemies of birds, and averred that the bush or tree rat was even more destructive than his grey relative.

The total number of stray cats in New Zealand exceeds the tally of human population, and is reputed to account for some ten million birds per year. One miner stated his cat caught on an average a hundred parakeets during the year. In towns they prey upon small native birds which show a tendency to adapt themselves to urban conditions. They are deliberately released in lonely places by owners anxious to be rid of them, and are imported by station owners keen to combat the rabbit pest. Others are abandoned by occupants of public works and other camps, remaining behind to live on native birds when man has passed on. The hedgehog of late seems to be taking its place in the ranks of destruction, accounting for countless pipits', skylarks', pheasants' and quails' eggs. In 1943 acclimatisation societies paid a bounty of sixpence per snout.

Following on 'information' obtained from Australia that they would never become a nuisance, opossums were released, and in 1892 the black opossum was introduced by the Wellington Acclimatisation Society. The animals quickly became numerous, and in 1912 an agitation against their introduction was commenced, saying: 'Although opinion is divided as to whether opossums are destroying bird life through sucking eggs, there is no doubt whatever that the operations of opossum-trappers during the season just closed have resulted in the wholesale destruction of ground birds, particularly the kiwi.' The belief that native birds are being helped by catching their enemies in traps is offset by the fact that, where every bird killed constitutes a serious loss, rats, being prolific breeders, rapidly make up the loss of the numbers caught in traps. In 1924 A.S. Wilkinson described damage done to native forests in the Tararua Ranges by opossums: 'I have travelled down gullies for hours where *Melicytus*, *Fuchsia* and *Nothopanax* trees grow, without seeing a healthy specimen, most of them with all the leaves and young shoots eaten off and hundreds dead. As *Fuchsia* and *Melicytus* are much frequented by native pigeons it can easily be seen how much harm the opossums are doing.' Another menace—though seldom referred to—is the illegal use of cyanide poison by unscrupulous trappers. Cyanide, particularly in misty weather, spells death to any bird which alights upon a tree above the poison. ...

Shore birds of New Zealand succumbed rapidly to the advent of civilisation, and so few remain that visitors often comment upon the lack of life amidst scenes of great coastal beauty; the once plentiful feathered inhabitants have declined before the unceasing assaults of vandals and vermin, and even visiting waders are hard-pressed to exist in face of modern firearms and changing food conditions. Godwit arriving at the end of their tremendous journey from Siberia were at the mercy of sportsmen until their case was taken up by the Forest and Bird Protection Society of New Zealand and they were placed upon the list of protected birds. Until recently New Zealand had no sea-littoral sanctuary for waders and coastal birds. But in 1938 Cape Farewell Spit ... was declared a sanctuary, though devoid of proper supervision. On the whole, oceanic birds such as penguin, petrels and albatross have fared better than most species of native birds because they breed, for the most part, upon outlying islands away from the haunts of man. But of recent years the menace of oil has to be included in the list of their tribulations.

Swamp birds have likewise felt the full force of civilisation owing to their territory having been drained and completely altered. As, for instance, when Piako swamp of 90,000 acres was drained in 1880, and 30,000 acres of Waikato swampland opened up for sale. Moreover, property owners continually burn the rank vegetation that remains.

Guthrie-Smith in *Tutira* considers that cattle were responsible for the elimination of waterfowl because they trample in their nests, destroy food and shelter. For in a land of limited acres the figures stood at 4,533,032 cattle and 31,062,875 sheep in 1940. These figures explain why native birds have lost their old feeding grounds of berry-bearing trees on the plains, which are now farms, and have been driven back into poor beech country where they cannot live. A concrete example of the harm done by stock can be seen on Stephens Island, which, until 1880, was clothed with rich vegetation, the home of many of the rarest birds in the South Island. Today, eaten out by stock, the forest has disappeared —save for small patches of trees clinging desperately to a water- less waste. The Auckland Islands—approximately 116,000 acres— famous for their unique plant and bird life—some years ago were leased at 15 shillings a week for pasture purposes. Finally the Royal Society of New Zealand intervened to prevent these islands sharing the fate of the Campbell and Enderby Islands, where the plant life was destroyed in an attempt to utilise them as grazing runs. The ruin of the South Island can be traced to the grazing of sheep on high tussock country in the interior. The annual burning off of this grass culminated in its elimination and con- sequent wrecking of such areas for grazing or any other purpose. The soil carried down by floods, to silt up harbours and river beds before going out to sea, caused a Maori sage to comment: 'The white man is building up another continent out to sea.'

Whilst cattle do harm, their inroads are slow compared to those of browsing animals within the forest itself. The final blow to the fauna of New Zealand was dealt when deer were introduced. The effect has to be seen to be believed. They ate seedling trees, preventing the replacement of mature trees that die, and destroyed others by barking them. By eating shrubs they enabled the wind to enter forest; they killed ferns, sedges, and other small plants of the forest floor, thus lessening its water-holding capacity. By trampling they hardened the surface soil and promoted more rapid run-off of rainwater; they devoured the absorbent mosses of the sub-alpine forest; and by bruising the surface roots of the trees and shrubs made wounds in which spores of wood-destroying fungi found entrance. Unfortunately, the damage they were doing was not recognised until too late. In spite of the protests of nature-lovers and of farmers, whose land the deer invaded, until recently restrictions were placed upon the shooting of deer. By 1930, so urgent had the matter become that in that year a deer conference was held in Christchurch, under the auspices of the Minister of Internal Affairs. In 1934 the *Southland Star* published an article saying that red deer had advanced into virgin country of South Westland, leaving a trail of ruin behind them. The 1933–34 seasons had seen the extermination of thousands of beasts. Two hunters alone had killed 2,000 between them. In the vicinity of the Turnbull River country, Messrs Barton and Barra reported hundreds of thousands of trees, beeches and pine, in danger. The deer had eaten out all undergrowth and even the bark of trees. Above the bush line the covering of alpine flora, for which the region was once famous, had disappeared. Now, instead of bush-covered hillsides, huge shingle slips had appeared, filling the rivers. … In spite of frequent culling, deer still abound; and in 1943

Mr Noland, writing from Dunedin, reported the deer as being present in the grazing country of South Westland 'to an alarming extent'.

Though less in the public eye, goats do even more harm than deer to the native forest and are extremely numerous. On the Kermadec and Three Kings Islands they not only destroyed for ever plants which were unique, but reduced the vegetation as a whole to a point at which bird life must cease to exist. …

Most native fauna was wiped out by the destruction of its food and home, but certain species were deliberately attacked because they clashed with introduced fauna and fish. The chief sufferer has been the shag. Says Guthrie-Smith: 'I believe there is no part of New Zealand where the destruction of shags is not thought to be a righteous action. On our rivers and freshwater lakes they are believed to harm imported trout, and on our seas to be a menace to the fishermen's interest. On Tutira the trout are increasing fast, although in the lake and on the run I have allowed no shags to be molested. It may yet prove … that the indiscriminate slaughter of shags will turn out to be inimical to the very interests sought to be preserved.' An opinion expressed by other naturalists such as Mr Edgar Stead. Despite these authoritative utterances the destruction continues; although certain coastal shags—protected by law—do not enter rivers inland. It is, however, a well-known fact that trout, which feed upon the young, have ruined what was once a lucrative whitebait industry. In 1943 Mr Howe pointed out that, if nothing was done, what was once a huge industry would be extinct within twenty-five years. When the above is considered, together with the fact that trout feed upon aquatic larvae of insects that furnish the food for birds, it seems a high price to pay for a sport in which the majority of New Zealanders—who much enjoy whitebait—do not take part.

The kea (*Nestor notabilis*), an interesting member of the parrot family, has likewise clashed with man's interest. Much of its food on the high country has been removed, whilst the bird itself has been slaughtered wholesale because of its alleged attacks upon sheep. Between 1920 and 1938 the government paid a subsidy on 29,000 kea heads, many of which were taken in a country where the birds had never seen a sheep. Whilst certain keas may develop habits like sheep-worrying dogs, the majority are innocent of the offence. …

Among the causes which have led to the decrease of bird life in New Zealand, the menace of firearms ranks high. Hutton and Drummond state: 'By shooting the number of avocet, white heron, pigeon, ducks, swamp hens, quail, and other birds were greatly reduced. In some favoured spots the slaughter was terrible.' … A common recipe for soup was 'one kaka and fourteen pigeon.' … Dr Oliver cites an example where kakapo, which were formerly abundant in the Makarora valley, were cleared out by men employed in forming the road. Railway and road construction through virgin forest is always an acute source of danger to the fauna of New Zealand. When the main trunk railway was under construction pigeons suffered most severely. But the worst weapon against which birds have to contend with in modern time is the pea rifle, to be found wherever man goes, in camps, launches, and motor cars. They are, *par excellence*, the weapon of youth who, when older, might be less cruel towards the fauna. In 1923 the report of the Forest and Bird Protection Society ran: 'The dominant factor, judging from correspondents' reports, is the uncontrolled use of the shotgun and pea rifle.' Sometimes ignorance causes a protected bird to be shot, as for example when the protected brown teal of Auckland province was destroyed in mistake for black teal. Both are now on the protected list. But ducks have suffered so severely from loss of habitat and the deadliness of modern firearms that certain species are expected to die out unless more effective protection can be given them. Under 'miscellaneous causes' for the disappearance of bird life can be cited: taking them for pets and the shooting of birds, or robbing their nests by farmers or others on behalf of local bodies, because the birds are believed to be harmful. The encouragement of such practices by children largely militates against the teaching of wildlife conservation in schools.

Early in the history of New Zealand, settlers introduced birds which reminded them of their countryside, or had proved useful to them in the past. Thus, acclimatisation societies arose throughout the country, whose policy it was to establish songbirds, game birds and utilitarian species, and [they] became one of the major channels through which wildlife was controlled. Of the 130 species introduced for sport, sentiment or to destroy insect pests, fortunately only twenty-five to thirty have established themselves, and until proper investigations have been made it is impossible to assert with authority whether certain species do more harm than good. In areas whence native birds have retreated, they keep a check on insect life which otherwise would increase rapidly. Where they come into contact with native birds, they must compete with them to a certain extent for food and nesting sites. One serious aspect of the acclimatisation of the pheasant has been the wholesale destruction of the woodhen—which is a known egg-stealer, though a staunch enemy of rats. ...

The importance of our unique wildlife is taught in schools, but takes a very small place indeed in proportion to the emphasis laid upon other subjects; and in private life they learn no better. For, despite all that has been written and the unceasing effort of the Forest and Bird Protection Society of New Zealand in conjunction with other bodies anxious to help before it is too late, the public of New Zealand are, in the main, utterly ignorant or not sufficiently interested to care whether their unique fauna vanishes or not. Indifferent to the example of other countries, they continue to shoot native birds which are, with few exceptions, on the protected list; fire natural cover, hack down trees from mountainsides and hilltops, and ruin natural beauty with no compensating result whatever. The only effect is ruination of the hillsides, the plains below, and the disappearance of a 'sylvan Parthenon that God will plan but builds not twice' (The Hon. W. P. Reeves).

Nothing short of unceasing vigilance by means of an efficient system of wardens can prevent the laws of protection being broken by those whose vision is not clear enough to appreciate what the loss of the unique flora and fauna of New Zealand will mean to themselves and to future generations. …

And what of the birds themselves? Diminished larders can only support a certain number of birds, and it is useless to increase bird life by breeding unless the native food is correspondingly increased. In spite of this they are making astounding efforts to remain in existence. Pigeon, bellbirds, tuis, and several species of smaller birds have shown signs of adapting themselves to new environments. Others, less adaptable, have been driven back into the last remaining wilds of New Zealand, where from their mountain fastness they may stage a comeback, if given the opportunity. But, optimistic as one would wish to be, the odds are too many and too heavy against the native fauna. Action is urgently required unless it is to come too late. In many districts they and the forest have disappeared.

> 'Gone are the forest birds, arboreal things,
> Wild harmless hamadryad creatures, they
> Lived with their trees and died and passed away.'
> — The Hon. W. P. Reeves.

So one asks: By what means can New Zealand be educated from without, and the public conscience awakened from the sleep of ignorance and apathy to a sense of the responsibility this country owes to the rest of the world for the unique flora and fauna with which she has been entrusted?

From 'The Destruction of an Avian Paradise' by Pérrine Moncrieff in *Forest and Bird*, May–November 1949.

The Passing of the Forest:
A Lament for the Children of Tané

In her article 'The Destruction of an Avian Paradise', Pérrine Moncrieff quoted the last lines of William Pember Reeves' poem *The Passing of the Forest*. Reeves published this poem in 1898 in London, where he was working as agent-general for New Zealand—a position that later became high commissioner. He had already had a career as a member of parliament in New Zealand, and in London he published poetry, as well as political tracts and a short history of New Zealand.

By the end of the nineteenth century, vast areas of New Zealand's native forests had been cleared to create farmland for European settlers, a practice that destroyed the habitat of many native bird species.

All glory cannot vanish from the hills.
　Their strength remains, their stature of command
O'er shadowy valleys that cool twilight fills
　For wanderers weary in a faded land;
Refreshed when rain-clouds swell a thousand rills,
　Ancient of days in green old age they stand,
Though lost the beauty that became Man's prey
When from their flanks he stripped the woods away.

But thin their vesture now—the trembling grass
　Shivering and yielding as the breeze goes by,
Catching quick gleams and scudding shades that pass
　As running seas reflect a windy sky.
A kinglier garb their forest raiment was
　From crown to feet that clothed them royally,

Shielding the secrets of their streams from day
Ere the deep, sheltering woods were hewn away.

Well may these brooding, mutilated kings,
    Stripped of the robes that ages weaved, discrowned,
Draw down the clouds with soft-enfolding wings
    And white, aerial fleece to wrap them round,
To hide the scars that every season brings,
    The fire's black smirch, the landslip's gaping wound,
Well may they shroud their heads in mantle grey
Since from their brows the leaves were plucked away!

Gone is the forest's labyrinth of life,
    Its clambering, thrusting, clasping, throttling race,
Creeper with creeper, bush with bush at strife,
    Struggling in silence for a breathing space;
Below, a realm with tangled rankness rife,
    Aloft, tree columns in victorious grace.
Gone the dumb hosts in warfare dim; none stay;
Dense brake and stately trunk have passed away.

Gone are those gentle forest-haunting things,
    Eaters of honey, honey-sweet in song.
The tui and the bell-bird—he who rings
    That brief, rich music we would fain prolong,
Gone the woodpigeon's sudden whirr of wings,
    The daring robin all unused to wrong,
Ay, all the friendly friendless creatures. They
Lived with their trees and died and passed away.

Gone are the flowers. The kowhai like ripe corn,
    The frail convolvulus, a day-dream white,
And dim-hued passion-flowers for shadows born,

Wan orchids strange as ghosts of tropic night;
The blood-red rata strangling trees forlorn
 Or with exultant scarlet fiery bright
Painting the sombre gorges, and that fay
The starry clematis are all away!

Lost is the resinous, sharp scent of pines,
 Of wood fresh cut, clean-smelling for the hearth,
Of smoke from burning logs in wavering lines
 Softening the air with blue, of brown, damp earth
And dead trunks fallen among coiling vines,
 Slow-mouldering, moss-coated. Round the girth
Of the green land the wind brought vale and bay
Fragrance far-borne now faded all away.

Lost is the sense of noiseless sweet escape
 From dust of stony plain, from sun and gale,
When the feet tread where quiet shadows drape
 Dark stems with peace beneath a kindly veil.
No more the pleasant rustlings stir each shape,
 Creeping with whisperings that rise and fail
Through glimmering lace-work lit by chequered play
Of light that danced on moss now burned away.

Gone are the forest tracks where oft we rode
 Under the silvery fern fronds, climbing slow
Through long green tunnels, while hot noontide glowed
 And glittered on the tree-tops far below.
There in the stillness of the mountain road
 We just could hear the valley river flow
With dreamy murmur through the slumbering day
Lulling the dark-browed woods now passed away.

Fanned by the dry, faint air that lightly blew
    We watched the shining gulfs in noonday sleep
Quivering between tall cliffs that taller grew
    Above the unseen torrent calling deep,
Till like a sword cleaving the foliage through
    The waterfall flashed foaming down the steep,
White, living water, cooling with its spray
Fresh plumes of curling fern now scorched away.

The axe bites deep. The rushing fire streams bright;
    Swift, beautiful and fierce it speeds for Man,
Nature's rough-handed foeman, keen to smite
    And mar the loveliness of ages. Scan
The blackened forest ruined in a night,
    The sylvan Parthenon that God will plan
But builds not twice. Ah, bitter price to pay
For Man's dominion—beauty swept away!

From *The Passing of the Forest and Other Verse* by William Pember Reeves: George Allen & Unwin Ltd, London, 1925.

How to Cook Paua

A.W. Baden Powell (1901–87), an enthusiastic conchologist, curated the Auckland Museum's mollusc collection and served as the museum's assistant director from 1936 to 1968. His classic work *Native Animals of New Zealand* was first published in 1947, and has been in print ever since. This handbook of New Zealand fauna includes simple descriptions and recognisable illustrations of more than four hundred species. Curiously for a natural history book, Powell's entry on the paua (*Haliotis iris*) includes cooking instructions.

[The paua] grows up to six inches in diameter, and is one of our most handsome shells. It is at once recognised by its large, oval, flattened shell, the row of holes along the back and wonderful internal lustre of opalescent greens and blues, with occasional fiery flashes. The shape of the paua is a special adaptation for clinging to flat surfaces of rock, after the manner of a limpet; the holes in the shell being for the purpose of expelling water used in the aeration of the gills. The paua is found at lowest spring tide level, and in deeper water, on rocky ground in open coastal situations. It is seldom exposed to view and the rough encrusted exterior of the shell renders it almost indistinguishable from its surroundings. Pauas cling to the rock with great suction, and a quick deft thrust with a broad thin bladed knife is necessary to prise them off. They favour deep low tidal rock pools, under sides of boulders, beneath ledges and in narrow channels and crevices in the rock. The best localities for the paua are Great Barrier Island, Wellington coast, Kaikoura, Stewart Island and Chatham Islands.

The paua animal has a considerable food value and is very palatable, provided the following rather drastic culinary preparations

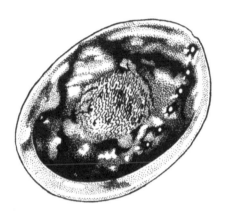

Paua, illustrated by
A.W. Baden Powell.

are attended to:—Remove the animal from its shell and discard all the soft parts, leaving only the tough foot and muscle, and taking care that a long white ribbon-like structure is removed from the mouth. This is the dental apparatus, which is studded with hundreds of hard, sharp, tiny teeth. If you do not like the black appearance of the animal, this coating will rub off with a coarse rag, leaving the flesh a dirty white or blue-grey colour. Next place the animal inside a cloth and pound it with a heavy piece of wood or the flat of a hammer, just sufficiently to relax the muscular tension. The paua is now rolled in flour or covered with batter and grilled for three minutes. Omit the pounding, or grill for more than three minutes and the paua becomes as tough as old leather.

Species related to our paua are highly esteemed in other countries, particularly in California, Japan and Guernsey Island. The paua was almost a staple food with the old-time Maori people, who used the shell also to great effect in their carvings and in the making of fishing spinners. Paua shell is now much sought after for the manufacture of trinkets, but as yet no serious attempt has been made to market the animal as food.

From *Native Animals of New Zealand* by A.W.B. Powell: The Unity Press, Auckland, 1947.

The Rediscovery of the Takahe

The first scientific description of the takahe, the world's largest living rail, was made after Walter Mantell (1820–1895), a government employee with a passionate interest in natural history, in 1849 acquired the skin of a bird that had been captured by some sealers in Fiordland. He sent the skin to his palaeontologist father, Dr Gideon Mantell, in London with this description of the bird's capture.

It was caught by a dog up a gully of the sound behind Resolution Island, Dusky Bay. When caught it made a great noise which he and his brother encouraged thinking to bring up its mate, but were afterwards told by the old natives that such noise was sure to drive the others away. As it was snowing and bitterly cold they did not continue the chase. The ground was covered with snow retaining tracks which they noticed and followed long before they saw the bird. Kept it alive on board the schooner for several days then killed and ate it. It was very good and as it was the first of its kind all hands had a taste for curiosity.

There were few other takahe sightings and only three more skins were collected. After a bird was taken from the shores of Lake Te Anau in 1898, no more were seen and the bird was presumed to be extinct. Fifty years later, however, Geoffrey Orbell, a medical doctor with a strong interest in natural history, discovered a takahe colony living in a glacial valley in the Murchison Range, 670 metres above Lake Te Anau. Joan Telfer (born 1926) and her fiancé Rex Watson accompanied Orbell on his expeditions to the colony, and Telfer published this account in 1949. At this time the takahe was called *Notornis mantelli.* Today it is known to be closely related to the pukeko, and is called *Porphyrio hochstetteri.*

Dr Orbell's interest in notornis began some thirty years ago. While looking through some old photographs belonging to his mother—a keen amateur photographer—his curiosity was aroused by a print of a bird mounted and displayed in a glass case. His mother explained that it was a picture of the notornis specimen in the Otago Museum, and that only four of the species had been found and naturalists thought that all the representatives had passed away with the extinct moa, eagle and swan. Dr Orbell's interest was stimulated and so he read everything he could find about notornis. As he grew older the call of the outdoors became very strong.

Frequent expeditions, such as deerstalking and pig hunting, led him to Beaumont in the Blue Mountains in Otago, where he met Mr Norman Murrell, the state forest warden. Mr Murrell, who was born and brought up at Lake Manapouri on the fringe of the notornis country, related stories of calls heard in the bush and of blurred tracks on sandy beaches.

An eagle eye was kept on all bird tracks while on a wapiti expedition up the north arm of Lake Te Anau, but only those of weka or kiwi were found. During this trip Mr Charles Evans, then ranger to the Southland Acclimatisation Society, told tales of tracks he had seen on a beach in Dusky Sound; of members of the crew dashing here and there on the beach trying to catch a bird the size of a goose, a bird with blue-green feathers and with the speed of a racehorse. He also told of tracks in the snow grass and droppings too big for any bird he knew.

From stories told round camp-fires, while stags roared challenges across high valleys, Dr Orbell gradually gathered the information which was finally to lead to the rediscovery of the notornis.

In 1935 Dr Orbell went to live and practise his profession in Southland and from there Te Anau became his hunting-ground. In this area Mr Ward Beer was acclimatisation society ranger.

He spoke of a large blue bird seen on the shores of Lake Ada and thought it a pukeko—still another link in the chain.

Also from a patient, Mrs Mackenzie, came another story of a large bird seen on the beach at Martin's Bay. Notornis? Notornis? This question was always there but rarely spoken of and then only jokingly or as bait to attract some story. Maps of Fiordland were studied during long evenings and wet days. These perusals, together with a careful study of the birds, gradually crystallised a certain fact. All stories were similar in one respect—the birds were seen or caught on beaches below bush line, and, as far as could be ascertained from books and other reports, they were noticed in wintertime mainly during years when very heavy snowfalls were experienced. Somewhere in the vast Fiordland area colonies of notornis must exist and the most ideal spot would be the parts least explored by man. Maori history suggested the existence of a lake (the Lake of the Friendless) and that its whereabouts should be beyond the western shores of Te Anau. Its position was verified from the air.

First expedition

In April last year a party—Dr G.B. Orbell, Mr S.R. Watson and Mr N.R. McCrostie—went through to investigate. Within four hours they reached the open tops on the edge of a terrific precipice. Some thousand feet below, lying at the end of a beautiful valley, was a lake, its tranquil waters disturbed only by an occasional paradise duck. Through field glasses deer were spotted grazing below.

The flat was reached by descending great overhanging bluffs and rocks and finally through a tongue of bush. Once down in the valley the going was not quite as easy as it looked from above. The ground was very marshy and every step was an effort. The billy was boiled and later the men separated, each confident he would return with the best head. Only one deer was shot, a rather poor fourteen-pointer. Dr Orbell was returning to meet his companions when he heard an unfamiliar bird-call—two long deep notes repeated twice. Curiously, when Rex and Neil returned they accused the doctor of whistling to them over a .303 cartridge case. By this time it was mid afternoon, too late to investigate further, so it was decided to return to the outlet of the lake.

In shallow water near the gorge footprints were first noticed, and a few feet away on the shore still more marks, obviously made by a bird or birds. The measurements were scratched on the stem of the doctor's pipe and on arrival in Invercargill were quickly placed on paper and sent to Professor Marples, of Dunedin. At the same time a duplicate set was compared with all available books. This comparison showed definite promise. The middle toe was four inches long, the right toe three and one-eighth inches, the left three and a quarter inches and the back toe one and a half inches.

Professor Marples wrote back that the tracks were too large for a notornis and were probably those of a white heron. Dr R.A. Falla, director of the Dominion Museum, Wellington, also could not reconcile them with any known bird tracks. Rex and Neil were very disappointed, but Dr Orbell realised that, to a man experienced in reading nature signs, a line of fresh tracks on a beach was of far more value than a diagram on a piece of paper could be to the authorities on the subject. In addition, the added knowledge that the Otago specimen was a young female led to the suspicion that the tracks of an adult male had been taken. This was subsequently proved to have been the case.

Second expedition

The winter of 1948 brought heavy falls of snow down to very low levels over the mountainous Fiordland district, making it impossible for anyone to venture into the country. Although nearly seven months had passed since the discovery of those footprints, notornis was not forgotten. On November 19 a party, including Dr Orbell, Neil McCrostie, Rex Watson and myself, left Invercargill to make a second attempt.

At 3.30 a.m. on Saturday we were on board the *Takatumu* moving up the lake. Breakfast was being prepared down below and served on deck at dawn. Lake Te Anau was perfectly calm and a most spectacular scene lay before us as the vivid lights of dawn crept through. Only the indigo mountains in the distance separated the flame-coloured sky from its reflections. The weather was ideal for our expedition. We dropped anchor and landed our packs on the western shores of the lake about 5.30 a.m. On this occasion we travelled very lightly, carrying lunch, cameras and fifty yards of fishing net.

The trek was long and tedious, up terraced slopes. Through dense beech forests, twined with lawyer and obstructed by windfalls and undergrowth tangled with a mass of roots, ferns and rotting trees, our progress was slow and checked on numerous occasions. After three and a half hours' climbing we finally emerged on to a clearing covered with snow grass and boxwood, and further on to snow grass only. When crossing this flat we noticed that the snow grass had been pulled out at intervals in the same manner as a swamp hen pulls reeds. These places also showed many droppings suggestive of a large bird. A few minutes later Dr Orbell dropped suddenly flat on the ground and motioned that we should follow suit. The three of us thought immediately of a deer. But it was not a deer. To quote the doctor: 'Either a pukeko or it!'

Peering excitedly through the long snow grass we saw a strange-looking bird strutting about on a small patch of swampy ground. The bright crayfish beak and dark head were unmistakable. Dr Orbell crawled slowly and carefully through the long

Neil McCrostie and Geoffrey Orbell holding live takahe at the small lake above the western shores of Lake Te Anau, November 1948. C-24407-½, Alexander Turnbull Library, Wellington

grass, applied a telephoto lens to his movie camera and commenced shooting. At this stage he signalled that another bird had appeared on the scene. We took the fifty yards of net out of the pack and proceeded to drag it into a semicircular position. When this task was completed we were ready for the attack, and the doctor indicated to Rex to stand at the opening of the net and help drive the birds into the net. During all this time the birds were making a very penetrating gulping noise. Apparently not in the least scared, they walked into the net of their own accord, and when we attempted to release them from it they clawed with their powerful feet, snapped with their strong beaks and screamed violently.

This commotion roused a third bird which we were unable to catch. He hovered at the edge of the bush and, at intervals, within close proximity for about an hour. All this seemed too good to be true. Only 9.30 a.m. and two notornis had been captured. Everything had gone according to plan and the birds had walked into the net with very little persuasion from us. We carried them over to the lake shore where there was a stretch of beach. Here we lightly tethered them to two stakes driven into the sand. The notornis were not unduly upset and quickly settled down. A few yards away we prepared lunch and looked with awe at our captives. …

Of the two birds one was smaller and appeared to be the female. The colouring of this bird was slightly more drab, her ear coverts were worn bare and her tail feathers were frayed, indicating that she had been sitting on a nest. The tail has the same twitch as seen in the pukeko, but the white under-feathers do not show so markedly. The beak of the notornis is very powerful— this can be vouched for by one member of the party who received a nasty bite. They were about the same size as a large table fowl and the feathers were of a beautiful silky texture.

Although they are flightless birds, the male we caught and measured had a wing-spread of 34 inches; however, the primaries and secondaries were very weak. Contrary to general opinion, the angle of the bird's back sloped downwards not upwards. The colouring was truly glorious—crayfish-red beak, legs and feet; navy blue head and breast, turquoise over the back; lower teal blue and finally tail feathers of olive green.

When we had completed our observations and the birds were released, we returned in what *Time* magazine so aptly describes as 'a state of ornithological ecstasy'.

Third expedition

Following the great interest aroused by the rediscovery of the takahe, another expedition was organised to make further studies of the birds in their natural habitat.

The flora and fauna of New Zealand and their preservation come under the government's Department of Internal Affairs and it was at that department's request that Dr Orbell acted as guide and leader of the expedition which left Te Anau on January 15, 1949.

The expedition was divided into an advance and main party. The advance party consisted of Dr Orbell and Messrs Woodrow and Wisely, of the Internal Affairs Department field staff. Their object was to find a suitable camping spot and to erect a camp to accommodate the party for a week. A base camp was set up in the bush on the western shores of Te Anau and from this point tents and supplies were packed into the notornis country.

At three thousand feet the forest has lost a lot of its grandeur, and with the valley floor so wet and marshy the finding of a camp site with suitable timber available for tent poles etc. was very difficult. A reasonably level spot was chosen among large rocks and boulders on the fringe of the bush. When the tents were erected

Dr Orbell and Mr Wisely set out to reconnoitre the lower end of the valley. Notornis signs were everywhere in evidence and very soon the characteristic 'thumping' sound and a pair of takahe were located within fifty yards of the spot where we had made our capture early in November. These birds were not disturbed and the doctor and Mr Wisely returned to camp with the knowledge that there seemed to be at least one pair in their neighbourhood for observation.

That same evening Mr Wisely returned to Lake Te Anau to act as guide for the main party due to arrive the next day. This second party comprised Dr Falla, whose task was to observe the habits of the birds and report to the Department of Internal Affairs, and Mr J. Sorensen, who conducted a botanical survey of the whole area in order that the feeding habits of notornis could be correlated with the flora of that particular valley.

With these two men came the other members of the original party, including Dr D. R. Jennings, Mr R. Weston and myself, bringing the total up to ten. Following a meal and a short rest, Dr Orbell conducted Dr Falla into the valley to give him his first glimpse of a living notornis. They were not disappointed for the birds were found in the same territory as they had been seen on the previous night.

The most interesting discovery that afternoon was a nest in which lay the remains of an egg and the dead and somewhat decayed body of a chick. It had been crushed just after hatching— probably early in December. The reason for the dead chick was not then evident. The shell was in many small pieces which were carefully collected in order that they might be fitted together to make at least a portion of the only notornis egg then actually known to be in existence. The chick appeared to have had the back of its skull crushed in. An answer was found the following day when the droppings of a stoat were seen in another empty nest. It seems probable that a stoat found the egg unprotected for a minute and, having broken the shell, was about to devour the chick when the parents returned. No doubt an adult notornis would be very effective against a stoat and the fact that the chick was left alone after the battle suggests that that particular stoat was unable to return to his kill.

The observations of the nest completed the day's work. At 3,500 feet, even in mid summer the nights are very cold, with icy winds blowing down the valleys from the snow-clad peaks at the head. Every available stitch of clothing was put on before we crawled into our sleeping-bags that night.

On the second day the whole party set out in line to search the valley floor. The sun was not up and the heavy dew on the snow grass, which was waist-high, made everyone uncomfortably wet. Mr Sorensen, the botanist, located a nest containing an adult bird. When she departed rapidly through the back entrance a small black chick ran, by way of the front door, directly to Dr Orbell's feet. This was a most interesting find. Dr Falla quickly made sketches, measurements and other notes. The youngster was very similar to a young European coot. It was covered by jet-black down and had very large purplish-brown legs and a black beak with a white tip. A notable feature was the presence of a very marked wing spur also characteristic of the rail family.

While the chick was being examined its mother was very disturbed. She came very close, making sounds indicating both alarm and annoyance. Other sounds appeared to be meant for the chick alone, for it replied with its own squeak, similar to that of any ordinary chicken. When measurements were completed and photographs had been taken of both chick and mother, our prize was placed on the ground to be met at once by its fond parent. Both hurried out of sight into the thick undergrowth at the edge of the bush.

During the rest of the morning many more nests, old and new, were found but all were empty of birds or eggs. Before a stop was made for lunch Dr Orbell parted two large clumps of snow grass and found a sitting bird. The surprise was, no doubt, mutual but the bird recovered first and made a hurried exit leaving in the nest one large egg. It was dull cream in colour with brown spots and faint mauve blotches. That the egg was fertile could be proved by listening to the movements of the embryo chick preparing to open its door to the world. For the benefit of the collector let me say that the egg measured 73.3 by 48.3 millimetres, and was photographed and replaced in the nest with as little disturbance as possible so that incubation might be continued.

The remainder of the day produced more unoccupied nests and a total of fourteen different birds were counted. Most of these gave calls which we had learned to associate with the presence of a chick.

On the following day several members of the party had to return to Invercargill and those remaining continued investigations to the north and to the south. To the south, Weston, Sorensen and Dr Orbell located another and smaller colony and saw one chick nearly half-fledged. Dr Falla was lucky enough to watch a bird feeding on snow grass. The diet of notornis being largely fibrous, the bird's droppings are readily distinguishable from all other bird signs and, once seen and identified, will prove the presence of notornis in any area although no birds may be seen.

The first day of the expedition proved the most interesting because the birds were relatively tame and could be approached very closely, but for the rest of the week they were very alert and could only be observed through field glasses.

However, the last afternoon Dr Orbell accompanied the botanist on his travels while the remainder of the party made the trip to the head of the valley in the hope of procuring some close-up pictures. The doctor sat in the snow grass at the edge of the beach on which the first photographs were taken. Rain was falling in the head of the valley and the light was very poor for photography when he heard a bird call on the opposite side of the lake. The call was immediately answered and he saw a bird come out on to the lake edge about one hundred yards away. Both birds continued to call back and forth and the second bird waded out into the lake and appeared about to swim to its mate. The depth of the water seemed to deter it and, still calling at intervals, it waded steadily in the doctor's direction.

The snow grass was a good hiding place, and after a few minutes' wading the takahe passed within three yards. In the poor light the beautiful green-blue feathers were still very noticeable and it was obvious that the bird did not mind getting the lower portions of its body wet. In this respect they are like the native pukeko or swamp hen which, although not web-footed, can comfortably swim quite large areas of calm water. The click of the shutter alarmed the takahe and it covered the remaining twenty yards to the opposite shore in record time amidst showers of spray.

This was the last view any of the party had of the birds, of which Dr Falla estimates there are at least one hundred. The next morning we returned to Te Anau, stopping on the way to collect a sack full of moa bones from recently discovered caves near the lake shore. These bones were those of several specimens of *Anomalopteryx*, the small bush moa of earlier times. The selection was made up mostly of pelvis, tibia and femora with a few ribs and vertebrae. The finding of moa bones was not, in itself, any remarkable discovery for these are common in all new caves in this area. But the fact that some of the bones were in a remarkably fresh state suggested the interesting possibility that not so many decades ago the notornis had a larger competitor for the available food supplies in that particular valley.

From a letter from Walter Mantell to Gideon Mantell, 19 January 1850; and 'Notornis Rediviva: A Mirror Nature Study' by Joan Telfer in *The Mirror: New Zealand's National Home Journal*, June 1949.

Takahe

After more than fifty years of supposed extinction, a small population of takahe was found living in the snow-covered tussock of a glacial valley high above Lake Te Anau. Poet Anna Jackson recalls the event.

For so long gone,
how strange

to find them again
not extinct

after all, small
families

passing on skills
for survival,

mother and father
teachers

in the intricacies
of tussock eating,

fern rooting,
and how

to hold food
with a foot.

Its beak a red
exclamation

mark, the takahe
shows how to find

the sweet core
through knowing

the tough exterior,
what to eat

through what
to leave:

watch me eat
the tussock core

so sweet
and so cold.

This is just to say,
takahe.

From *The Pastoral Kitchen* by Anna Jackson: Auckland University Press, Auckland, 2001.

Love-habits of the Yellow-eyed Penguin

To research his 1951 book *Sexual Behavior in Penguins*, Otago ornithologist Lance Richdale (1900–83) spent a decade studying the penguins of the Otago Peninsula. By the mid twentieth century, the study of fauna had moved on from the nineteenth-century focus on classification and laboratory work, and the study of birds in their natural environment was well-established. The work was intensive —Richdale accumulated some fifty weeks' fieldwork, often sleeping in a tent in isolated locations and enduring wild weather.

Here he describes some of the mating behaviour of the yellow-eyed penguin, one of the world's rarest penguins, whose habitat is now limited to the subantarctic islands, Stewart Island and a few mainland sites on the South Island's east coast.

Salute—As far as I have been able to ascertain, it is usually the male which adopts this attitude. A bird apparently interested in a female will stand from six to fifteen feet away with his back to her and … will seem to be quite indifferent to her presence. He must, however, be fully conscious of her, for suddenly he turns around and, with neck arched, with beak nearly touching the ground, and with flippers pushed out stiffly in front of him, he rapidly walks up to and past her with quaint little steps, stops with his back to her, and thrusts his beak straight up to the sky, with his neck fully stretched, and with his flippers still thrust out in front. Reaching to his full height he maintains this position for about five seconds before slowly lowering and turning the head, almost simultaneously looking over one shoulder as if to gauge the effect. It should be noted that when the head is lowered, it is accomplished by the contraction of the neck, the bill still pointing upwards, and the latter not being lowered till the bird slowly turns his head to look over his shoulder. Even then the bill is lowered only slowly.

Frequently, I have seen the demonstration confined to a single 'salute'. On other occasions, the male will again walk past the female, but not so far, and repeat the procedure. After that the pair may start to preen each other, and love-habits in varying degrees may follow. ...

The 'salute' may be used by one or more curious juveniles which congregate in characteristic fashion around a pair indulging in love-habits on the landing-ground. First one and then another will walk quickly past not only the male, but the female as well, and 'salute' until pecked off the premises by the irate male. ...

Sheepish Look—The bird stands up off its tarsi but does not stretch itself up to its full height as in the 'salute' or the 'full trumpet.' The flippers are pushed out stiffly straight forward, at an angle of 45° to the ground, but it is the appearance of the head which characterises the attitude. The neck, instead of lying normally contracted between the shoulders, is half stretched upwards, but not fully as in the 'salute'. The head is then hung so that the bill, pointing downwards, lies almost parallel to the foreneck. Frequently the head is held to one side ... while the sheepishness of the whole attitude is accentuated by the wide-opened staring eyes, and the raising of the feathers of the crown and forehead. ...

The 'sheepish look,' which is usually, though by no means always, adopted by the female, is very common after the birds remain ashore in the daytime in August, till about the middle of the incubation period. It may occur at any time when a bird returns to its mate at the nest, though one is not certain to see it included in the elaborate ritual that takes place during the change of guard. In the early winter months it may be witnessed when the birds perform love-habits either at the nest site or on the landing ground. From the end of July to the beginning of the pre-egg stage in August, it becomes increasingly common. Finally, it occurs in the period of moult, particularly between mated pairs and pairs 'keeping company'.

Throb—This action, beginning with rapid pulsating movements of the skin and feathers at the base of the neck and top of the breast, is really a further development of the 'sheepish look'. The bill, raised to the sky at an angle of 45° and sometimes a little further, is opened slightly, vibrated very slowly, and a noise resembling a series of chuckles issues from the base of the throat. The crown is raised slightly.

The 'throb' is common to both sexes and is frequently heard at irregular intervals when two birds of a pair, or a mated pair, are together. It may signify the commencement of a bout of intense emotional activity, or it may be the final act after intense activity before the pair subsides to a spell of passiveness. On other occasions, the periods of inactivity may be broken for a few moments by 'throbs' only, with no behaviour of greater intensity. …

Open-yell—On some occasions, a bird, without any preliminary actions, breaks into what I term the 'open-yell'. Standing off its tarsi, it leans forward with the body at least 45° to the ground, with the neck stretched out in a similar line, and with the head bent slightly forward off that line. The mandibles are then opened as far as possible, with the eyes staring to their fullest extent. As it lunges forward, the bird emits a yell at the top of its voice. The flippers are sloped down at 45° from the shoulders and are halfway between the 'arms forward' and 'arms sideways' positions.

This love-habit does not seem to have social value. In pair-formation it does not occur until the members of a pair are well disposed towards each other, and is employed mostly during the 'welcome' ceremony, after one bird has returned from the sea to find the other at the nest site.

The main function of the 'open-yell', as a love-habit, is within the family group and as such has family value. A bird will break into the 'open-yell' when, on the nest, it suddenly sees its partner returning from fishing … when it performs part of the 'welcome' ceremony … and when it has returned from the sea, at the post-guard stage, to observe the chicks issuing from their hiding-place clamouring for food.

Half Trumpet—This is the commonest behaviour of all the love-habit activities, and consists of 'throbs', and what appear to be half-hearted trumpets. Standing off its tarsi, the bird leans forward slightly, with its head still a little further forward. The mandibles are half open, emitting fairly loud continuous noises, which are difficult to define, while, at the same time, 'throbbing' occurs at the base of the neck. The flippers are pushed forward while hanging down at an angle of 45° from the shoulders, with the inner surfaces facing each other. …

The 'half trumpet' may be heard at almost any time when a mated pair, and sometimes when two birds 'keeping company', are found together. It is part of the 'welcome' ceremony, and will last for some considerable time after that is finished. In the period of moult, a pair or mated pair will suddenly break into this activity for a few moments, standing so that the bills almost touch each other. At the pre-egg stage, and whenever any love-habits are taking place in the winter, when a pair or mated pair happens to stay ashore for the day, it is very common. For example, during one important episode on record, two birds continued the 'half trumpet' behaviour at frequent intervals nearly all day…

Full Trumpet—A bird seldom breaks directly into the 'full trumpet,' for it is generally the culminating point of one or more of the preceding types of behaviour. Standing only on its toes with the tarsi and legs erect and in line, with the tail and underparts quite clear of the ground, with the flippers thrown stiffly forward at the 'arms forward' position, with the head thrust straight towards the sky, with the line from the chin through the breast to the vent forming the arc of a circle, and with the line running from the back of the head to the tail a definite hollow, the bird opens its mandibles, and the vast volume of trilling sound that issues forth is fantastically suggestive of the tremolo of giant crickets, and is musical to a degree. All this time, with the mandibles vibrating rapidly, the feathers and skin at the base of the neck and top of the breast, too, are pulsing vigorously in and out. A near view shows that the muscles from each shoulder down the back to the tail are contracting in rapid rhythm. …

Arms Act—I first witnessed this performance at 4.57 p.m. on September 4, 1937, two weeks before the first egg appeared, on the landing-ground, where the male was practising it on the female. The former had already given the impression that he was searching for a mate, when he suddenly went up to the female, whose back was to him. He pushed his breast up against her back, at the same time putting the front of his neck on the back of hers, and pressed heavily. All this time the male's flippers, vibrating rapidly, were one on each side of the female and protruding in front of her. It appeared as if he were trying to 'put his arms around her'. The female submitted to this treatment for a while and then moved away.

When I first saw this action it puzzled me very much, and continued to do so in subsequent seasons, for I had not seen a repetition of the performance in the yellow-eyed penguin. On March 31, 1939, however, I observed similar rapid movements of the flippers and the pushing of a male's neck on to a female's in the erect-crested penguin. As the female was prostrate on the nest and had obviously been stimulated by the male, it was apparent that this was the beginning of behaviour leading to coition, and my thoughts immediately reverted to the yellow-eyed penguin pair on September 4, 1937. It should be remembered that on my many visits to the nests of the yellow-eyed penguin in the ten years when I observed them, I never witnessed coition, although I had been present at night as well as in the daytime.

In January 1940, I several times saw attempts to copulate both by the male and the female erect-crested penguins, scenes which convinced me that the male yellow-eyed penguin was trying to persuade the female to copulate with him on the night in question.

From *Sexual Behavior in Penguins* by L.E. Richdale: University of Kansas Press, Lawrence, 1951.

The Third Man
of the Double Helix

Maurice Wilkins (1916–2004), James Watson and Francis Crick were awarded the 1962 Nobel Prize in Physiology or Medicine for their work on the structure of nucleic acids, including the discovery of the double-helix structure of the DNA molecule.

Wilkins was born in 1916 in the north Wairarapa town of Pongaroa. His family soon moved to Wellington, where they lived until Wilkins was six years old and they moved to England. In 1940, Wilkins completed a PhD in physics, and during the Second World War he worked first on the development of radar screens and then on the separation of uranium isotopes for use in nuclear weapons. Wilkins was horrified by the subsequent use of nuclear weapons on Japan in 1945—he became a lifelong supporter of the Campaign for Nuclear Disarmament and turned from physics to biophysics.

At the start of this narrative from Wilkins' 2003 autobiography *The Third Man of the Double Helix*, he is assistant director of the biophysics laboratory at King's College, London, under the leadership of physicist John Randall. It is 1950 and Wilkins is beginning work on the structure of DNA. He would later describe the scientific community as 'still living in the belief that DNA was little more than an accessory to the proteins that were the basic elements of genes.' In particular, Wilkins said of his friend Francis Crick, a physics graduate, that he 'thought I was wasting my time on DNA' and 'could not understand why I did not concentrate on something useful.' However, Wilkins' view was that 'if DNA really was the stuff of genes, then that was where my work should concentrate.'

As the story begins, Wilkins has just obtained a glass phial of DNA carefully extracted from the thymus glands of calves by the Swiss biochemist Rudolph Signer.[*]

Signer's DNA looked like a bundle of tiny white threads when it was dry. I thought that in a living cell the DNA would be wet, so I decided to keep the DNA moist. When it was wet, it formed a sticky blob that … looked like snot. In order to study the structure of the DNA molecules—or at least to find the way in which the chemical groups were aligned in them—I took the damp DNA, smeared it into thin sheets and used one of our microscopes to measure how polarised ultraviolet light was absorbed in the sheets. The Signer DNA was very different from other DNA I had used. I was intrigued to find that the sticky blob of DNA seemed to dislike forming sheets, and instead tended to form very long thin fibres like those in a spider's web. When I looked carefully at the fibres with a microscope, I saw that they were extraordinarily uniform and transparent—there was a perfection about them which led me to a strong intuition that the arrangement of the molecules in the fibres might form a regular pattern. If the organisation within the fibres were regular, then they were crystalline, and could be examined by X-ray diffraction. … If I was right, the fibres should give sharp X-ray diffraction patterns, and that might tell us very much more about the molecular structure of the DNA than was possible by using light microscopes. …

I was quite right in thinking that fibres of Signer DNA would help us to understand DNA structure. But I soon found a very peculiar thing about the fibres. They did not stretch uniformly.

---

[*] Other accounts of the race to discover the structure of DNA are found in James Watson's book *The Double Helix: A Personal Account of the Discovery of the Structure of DNA*: Weidenfeld & Nicholson, London, 1968, and Brenda Maddox's biography, *Rosalind Franklin: The Dark Lady of DNA*: Harper Collins, London, 2002.

When I applied tension to them they suddenly gave way, and a long neck pulled out until the fibre was almost twice as long. …

It was not easy for us to follow up the invitation the DNA seemed to be offering … the only X-ray equipment we had was an ordinary diffraction camera designed for routine study of single crystals that were much larger than our tiny fibres. Indeed, Randall seemed to have forgotten about X-ray work until 1950, when he asked his student Raymond Gosling to take some patterns 'to complement the studies' of sperm he was making with his electron microscope. I was friendly with Raymond and had tried to help him because he only got poor diffraction patterns from the sperm. He asked me for some Signer DNA for comparison, but smeared lumps of DNA only gave diffuse patterns. He was very glad to try DNA fibres instead, but the fibres were so thin that they would diffract X-rays only very little. I therefore made a little frame of tungsten wire … and after a struggle I managed, with watch-makers' tweezers and quick-setting glue, to stick a dozen or so fibres fairly parallel across the frame. I was not very patient—Astbury had once got a research student to line up several thousand tiny chromosomes for X-ray study, but their results were not encouraging. Our results, however, from the dozen DNA fibres on the wire frame were very promising.

It was very fortunate indeed that Randall had learnt … that weak diffraction could be seen much better when the air in the camera, which scattered the X-rays, was replaced by the very light gas hydrogen. That was a very important new tactic, but making the camera hold in the hydrogen was not at all easy. It was impossible to seal the bottom of the camera, but we sealed the top with Plasticine and fed hydrogen in at the top, hoping that, being very light, it would displace the air downwards … Another difficulty was that the X-rays entered the camera at the side through a tube that had to fit loosely so that its position could be adjusted. But I found that part of a condom made a good flexible seal (and that seemed to amuse people). Our apparatus could be seen either as a real bodge-up or a brilliant improvisation. The important thing was to get some results.

The key factor leading to our success was the feeling, which I had picked up from working with live tissue culture, that DNA should not be treated as though it were inert and non-living. DNA was a sensitive part of living things, and its structure would depend on how it interacted with its environment. While we bombarded the DNA with destructive X-rays we should at least do our best to keep it, so to speak, in as healthy a condition as possible. Signer had taken great care in preparing his DNA and we should expose it to the gentlest treatment possible. I had noted that when DNA fibres were kept in moist air they swelled and the ultraviolet-light-absorbing groups lined up to the greatest extent. I remembered too, from my student days, that Bernal* had obtained the first really sharp X-ray diffraction patterns from protein crystals by keeping them in the 'mother liquor' out of which they had grown—which made it sound almost as though they were alive. ... Such thoughts led me to believe that the fibres being X-rayed should be carefully preserved in a moist atmosphere of hydrogen. They should certainly not be allowed to dry or be treated with alcohol as some workers had done—all of them had failed to get the results they had hoped for.

Keeping these points in mind, Raymond and I very soon obtained diffraction patterns of DNA that were much sharper and more detailed than any before. X-ray diffraction pictures are not like medical X-rays, in which one's bones are clear for all to see: instead, they tend to show patterns of spots from which the

---

* John Desmond Bernal, one of the 1930s pioneers of X-ray crystallography.

internal structure of the subject can be inferred by a lengthy sequence of calculations. But the more regular the structure, the sharper the diffraction pattern will be. The pictures Raymond and I took consisted of dozens of well-defined spots on a clear background, which was a very good result: it showed us clearly, for the first time, that DNA was truly crystalline, and we could be very hopeful that its structure could be derived from X-ray patterns. The great community spirit and cooperation in our lab had produced this valuable result. I had discovered the extra-ordinary properties of the Signer DNA fibres; Raymond was building up experience of X-ray diffraction; and Randall had recommended replacing the air in the camera with hydrogen to make the patterns much clearer. Thanks to all these contributions, Signer's excellent pristine DNA was, so to speak, shouting at us, 'Look how regular I am!' On the other hand, we knew genes had to be very complicated and therefore DNA had to be compli-cated. That paradox presented us with a very important challenge in our thinking about the structure and function of DNA.

By 1950, X-ray diffraction had only provided fibre structures of fairly simple polymers. No fibre as complicated as DNA had been solved, and there was almost nothing to guide us. ...

One advance at that time was made by Raymond, who, with our colleague the physicist Alec Stokes, measured the position of all the spots on the DNA pattern and deduced that the DNA molecules were in a monoclinic crystalline arrangement (that is, skewed in one direction only). Raymond confirmed that the molecules packed together along the length of the fibre as cylin-ders, like a stack of pipes, as I had inferred. Stokes then came up with an enormously important idea. He had experience of diffraction from zigzag and helical structures when we had been studying virus crystals. I knew that chromosomes, which were largely DNA, were often spiralised, that is they had well-defined helical forms. Stokes, pondering on the DNA pattern, noticed that there was no diffraction in directions on, or close to, the fibre direction, and he realised that this absence of diffraction suggested that DNA might be helical. Knowing about chromosomes, I was very interested in Stokes' helical idea. We were so much at sea with the problem of interpreting the DNA pattern that we felt it was very important to consider any simplifying hypothesis, and the helical idea seemed very useful indeed.

While we were in some respects uncertain about how to proceed, we were very clear that we must pass out of the stage of preliminary experiment with a borrowed generator and a 'bodged-up' camera. Being in a physics department, we paid special attention to having good physical equipment. ... We considered a rotating anode X-ray generator to give a very powerful beam, but decided instead to use a new, fine-focus X-ray tube ... which was designed to concentrate X-rays on very small specimens. I felt that we should work with micro-specimens consisting of only one fibre, or only a small number, selected for uniformity and perfection of crystallinity using a polarising microscope. A special micro-camera would be needed. In that way we would be much better equipped for DNA work than most long-established X-ray labs. ...

I also thought that we needed more staff on DNA, especially an experienced X-ray worker. By the summer of 1950 I knew that Randall had appointed a new X-ray specialist to work on protein solutions. Her name was Rosalind Franklin, and she had been working in Paris studying the structure of coals. ...

Stokes had got the new fine-focus X-ray source working by the autumn of 1950, but Rosalind, as the incoming experimental expert, was to carry out our plan of attaching to the new source a camera suitable for small fibres of DNA, or even for a single, selected fibre. At first, over the range of swelling I used, the increase in fibre diameter and length happened to be the same, and I thought that the swelling might not correspond to changes in molecular structure. ... Later, I increased the hydration and found that the fibre diameter increased greatly as more water went between the molecules, but the length of the fibres would not increase beyond about 20 or 30%. When I had only preliminary humidity data for DNA in our adapted camera, I was led in the wrong direction by the swelling results and I began to push the idea that DNA consisted of only one helical chain (which would have been sufficient to contain all the genes). Rosalind was able to demolish that idea as soon as she began using a proper camera. She then showed that the lengthening of the DNA fibres was the same as the increase in the periodicity of the diffraction pattern. Neither she nor I nor Stokes realised how the lengthening of the DNA molecule and of the whole fibre were related, though with hindsight it all seems so obvious. The change of fibre length resulted from the helices of DNA partially uncoiling and lengthening. When that was recognised, it would turn out to be an important step towards the DNA structure. ...

> On a conference trip to Naples, Wilkins collected sperm from *Sepia*, a member of the squid family, whose sperm heads are composed largely of DNA molecules. He also met James (Jim) Watson, a young American scientist who had recently completed his PhD on the effect of X-rays on the reproduction of bacteriophage, viruses that infect bacteria.

The prospects of our DNA X-ray work brightened as summer began. ... Back in London, it was good that Rosalind had begun reconstructing our new X-ray equipment. While waiting for that, I made do with the old set-up we had used a year before for our breakthrough in understanding the crystallinity of DNA. I began thinking about the basic biological importance of the DNA pattern, and wondered if the structure of Signer DNA (which was made from calf thymus glands, sold as sweetbreads by butchers) was basically the same as that of DNA in other living things. Was there just one universal gene structure? If so, our sharp Signer pattern could have a very wide (and big) significance. Mary Nicholls in our lab had extracted DNA from humans and herrings as well as calves, and I decided to compare the diffraction from her DNA with that from the DNA from the *Sepia* sperm that I had brought back from Naples. …

I was very encouraged to find that DNA from the various sources all gave patterns that were basically the same. That could be very significant biologically; and it could mean that structure work would be much simpler than with proteins, which had very many different kinds of structure from one species to another. Also, although the patterns were very diffuse and rather different from the sharp Signer pattern, it was exciting that they all showed fairly clearly a central X. That feature, as Stokes had urged, was a strong sign of a helix. … I discussed these results in a talk I gave in Cambridge in July 1951, at a meeting of protein X-ray workers organised by Max Perutz … Bill Seeds came with me, and sat in the back row with Francis Crick while I described how it seemed that all DNA had the same unique, universal structure which was twisted regularly into a helix. We were thinking in a new way—in terms of the overall shape of the molecule … The angle of the X on the patterns was about 45°, and that was the angle of ascent of the helix (like the slope of a spiral staircase). The sharp Signer pattern gave the diameter of the staircase as 20Å, and the distance ascended in one turn was about 27Å (Å stands for ångströms—there are one hundred million ångströms in one millimetre). The *Sepia* sperm pattern showed that we were looking at the structure of living DNA. Everything fitted together; and if DNA really had the simplicity of a helix it might enormously simplify the problem of finding the complete structure. ...

My talk provided a coherent story. Jerry Oster said that a scientific lecture needed to tell a story, and, like a theatre performance, the clapping at the end was a measure of its success. ... I had an extra satisfaction in giving my talk: 15 years earlier, as a first-year student, I had listened in that same lecture theatre to the great Rutherford giving lectures on physics.

Wilkins then travelled to the United States to attend the 1951 Gordon Conference on Nucleic Acids.

Meeting the biochemist Erwin Chargaff, from Columbia University in New York, was the most important event for me. He was very friendly and helpful, and seemed very interested in our X-ray studies of DNA. He had analysed the nitrogen bases in DNA, and had found that of the four bases, guanine was present in the same quantity as cytosine, and adenine in the same quantity as thymine. But he did not suggest that these 1:1 ratios were achieved by pairing, which would be the accepted explanation a few years later. ...

I returned from the USA very cheerful, very positive and possibly a bit cocky. Important talks with leading DNA people, much interest in what we were doing; it had been very good for my ego. It had been exciting flying the Atlantic, and after a week I would be flying off again, to Berlin ... Then I had to go with Randall and others in our lab to a conference in Stockholm, which focused on microscopy and other biophysics techniques...

Back in London ... Rosalind had discovered a somewhat different type of X-ray diffraction pattern when the DNA was very wet—about 92% humidity. She showed me the new pattern, which we called B-DNA, which was similar to but much clearer than those we had made before. B-DNA looked set to lend itself to fruitful analysis. The DNA that gave the crystalline pattern, which we now called A-DNA, was not as wet as Raymond and I had thought—the humidity was only 75%. Clearly, Rosalind said, the DNA molecules could exist in two different structures, and changing the humidity from 75% to 92% caused the structure to change from A to B. It was clearly an important discovery. But she put so much emphasis on what I had got wrong that I could barely take any pleasure in it. ...

Another aspect of this situation is that it illustrates that scientists often become very strongly attached to their research. I did, and I think Rosalind did too. What the poet Coleridge said is still true: the scientist loves the material on which he or she works. For science to function for the benefit of society, this point should be kept in mind. Mountain climbers risk their lives to reach their goal, and scientists have strong feelings too. A single climber cannot reach the top of Everest without support from others. Centuries ago a single scientist could make much progress on his own, but today working in a group is almost essential, and so skill in building relationships is increasingly important.

For us, the question remained: how could our X-ray work be organised so that we could proceed in a reasonably friendly manner? ... I was still very excited by Chargaff's work, and I suggested that I should study the DNA that formed fibres that Chargaff had given me; and that Rosalind should continue with Signer DNA. I had designed a new camera that could be sealed up like a microcamera, but would give a somewhat larger X-ray pattern. Our excellent mechanics could make my camera in a few weeks. As a temporary matter I was prepared to use the old Raymax X-ray set that Raymond and I had used a year before. Rosalind had exclusive use of our new fine-focus X-ray tube. She agreed that I should try the Chargaff DNA while she and

Raymond continued working with the Signer. Why did I not ask for a share of the Signer DNA? That was a mistake on my part. Signer had, after all, handed it to me in May 1950. … My self-effacing manner was probably not productive, and may not have impressed Rosalind. Randall, as head of the lab, should have helped in theory: he called Rosalind and me to a meeting and suggested that Rosalind continue to work on the crystalline A pattern, and I could work on the B pattern. That seemed sensible, but his manner of telling us did not help matters: he said he wanted to be fair to both of us, and that made me feel like a naughty child.

Our situation was much altered in the summer of 1951, when the great chemist Linus Pauling published a paper on the protein α-helix [alpha helix]. Pauling was working at the California Institute of Technology, and had been encouraged in the early 1930s by a grant from the Rockefeller Foundation to work on the structure of biological molecules, especially proteins such as haemoglobin and antibodies. … In May 1951, he wrote seven papers giving the atomic arrangement within the molecular structure of many proteins, including the structure of the most important fundamental form of protein: a helical chain known as the α-helix.

Pauling had made this important advance in a fundamentally new way, by both thinking about the known details of inter-atomic bonding, and making large-scale models of possible molecular structures. … Clearly model-building had been put on a new scientific basis, and many scientists were stimulated to try it. I encouraged Bruce Fraser, in our lab, to try out his ideas in a model. … A new way of exploring had come into being.

When I had finished reading Pauling's α-helix paper I was puzzled that he seemed not to have a way of calculating X-ray diffraction from the structure. I discussed this with Stokes, and our cooperation on this question was very fruitful: the next day he produced a Bessel function calculation of diffraction from a helix. He had worked it out on the train to London from his home at Welwyn Garden City. … The calculations confirmed Stokes' thoughts of a year earlier, that diffraction near the fibre direction would be very weak if DNA were helical. What was really exciting was how well the plot of calculated intensities corresponded with Rosalind's new B diffraction pattern. Stokes and I had a real feeling of scientific uplift, and we felt we must share this with Rosalind. We did not wait to think about it, but hurried along the corridor to show Rosalind how very in-formative her pattern had become. When we came to the room where Rosalind was working, all I can remember was seeing her standing in the room listening to us trying to explain what we saw as very important good news. Then suddenly she angrily exclaimed: 'How dare you interpret my results!' We were flabber-gasted, and did not know what to say. It was the only time I ever saw Rosalind lose her temper. …

The clearest memory I have of the whole incident is of seeing a continuous straight grey line on the diffraction pattern photo-graph. The intensity variation along one line corresponded to one of the waves of the Bessel function. Stokes had a clear memory of Rosalind's B pattern, and the theoretical results matched very well. …

In October 1951 … a laboratory colloquium was set up to survey the DNA X-ray work in our lab. We had planned this as an opportunity for open discussion; and when Jim Watson asked if he could attend we did not hesitate to say yes. In his new post at the Cavendish Laboratory in Cambridge, he had met Francis Crick, and they had formed a powerful team. They shared en-thusiasm for Pauling's new model-building and for Stokes' Bessel function calculation of diffraction from helical molecules.

Francis's interest in X-ray diffraction as a fundamental biological tool complemented Jim's enthusiasm for bacteriophage genetics. Together they formed a very effective complementary pair for the study of DNA structure.

I opened the colloquium by repeating the talk I had given at Perutz's meeting in Cambridge in July. I went through the X-ray evidence that DNA from a wide range of species gave a basically similar 'cross-ways' X-ray pattern indicating the same helical structure. I regretted that I had little new to say, but I probably mentioned Chargaff's important results. I didn't discuss the new helical evidence. Stokes then described his Bessel function description of diffraction from a helical structure, but I do not think he attempted to link his work with Rosalind's new B pattern. Rosalind then gave a first-class account of various likely aspects of DNA structure. She clearly presented reasons why the phosphate groups should be on the outside of the molecule, and the importance of understanding the role of water in DNA structures A and B. ...

Apart from tension over helices, our lab was at that time very well placed indeed to discover the structure of DNA. We had masses of new X-ray data. Rosalind was well versed in X-ray techniques and understood the physical chemistry of DNA. Stokes was a master of diffraction theory. We knew the X-ray workers at Birkbeck College who were finding the structures of the component parts of DNA. A Norwegian research student at Birkbeck, Sven Furberg, had sent us his PhD thesis about very important X-ray work on DNA. We were in touch with the various relevant chemical and optical groups in our lab, especially Bill Price's infrared spectroscopic group. Jim Watson stimulated us with new ideas from the USA, Francis Crick was always thought-provoking, and we were inspired by Pauling's great success with the α-helix. All we had to do was to get on with the job.

The strength of our position was illustrated by a model of DNA that Bruce Fraser built. ... Fraser worked in the room next to mine and, soon after the colloquium, he appeared at my door with a mysterious smile and beckoned me silently. Following him into his room, I saw that he had built a helical model of DNA. Bruce had done a good job: the model was very interesting. The three helical chains had the right pitch, diameter and angle, and were linked together by hydrogen bonds between the flat bases which were stacked on each other in the middle of the model. But the three chains were equally spaced, and that was contradicted by the X-ray diffractions. There were also basic difficulties with the hydrogen bonding between the bases: the bonds could only exist for special groups of three bases. The structure did not fit with Chargaff's 1:1 base ratios.

The model was designed to follow estimates of the energy of binding between stacked bases, of hydrogen bonds linking bases and of forces between phosphate groups. ... The model represented well the general state of thinking in our lab. ... But the difficulty was that we just did not know what to do with the three helices. We found ourselves completely stuck. There seemed to be no use at all in trying to build more models unless we could find some idea to guide us; and we could not find that idea. Thinking that there were three chains had completely stopped us in our tracks.

> Wilkins' pro-helix approach, coupled with a difficult working relationship with Rosalind Franklin, who had presented evidence against the DNA molecule being a helix, stalled the team, who were not working effectively together.

Two weeks after our colloquium at King's, I was surprised by Francis Crick, who telephoned me to say that he and Jim Watson had built a model of DNA and would I like to come to Cambridge to see it? Rushing round the labs, I collected Rosalind, Raymond, Bill Seeds and Bruce Fraser, and we were soon filling a compartment on a train to Cambridge. We knew that Francis and Jim were very bright, and we wondered what they had come up with.

We were surprised to find the model disappointing. According to our thinking it was completely inside out, with the helical regularity in its three chains being established by the phosphate groups held together along the helix axis. But magnesium ions (which we had never heard of in DNA) were needed to hold the phosphates together. On the outside of the helix, the bases seemed to flop about without being stacked on top of each other, or stabilised in any way. When Francis saw that we were not impressed, he said the model was based on a low water content that Jim had misunderstood at our colloquium. We did not take the model seriously; but if we had known that Pauling was to produce a rather similar model in a year's time, we might have been more patient and respectful. Bruce spoke about his model, but Francis and Jim were preoccupied with their own misadventure.

However, in spite of Francis and Jim's failure, which did not seem to lead anywhere, we had a strong feeling that what they had done was only a small beginning, and that they intended to press on. I had seen enough of them to expect that they could be very serious competitors if there was to be a race for the DNA structure. The idea that all of us at Cambridge and King's should work together on DNA did not occur to us: King's had opened up the field and we felt we should be free to explore the possibilities that we had created. On the other hand, we were two MRC* labs managed and funded by the same agency, and it seemed silly for us to compete with each other. There certainly were some feelings of rivalry between the two labs. ...

---

* Medical Research Council.

If Francis and Jim's helical model of DNA had more adequately demonstrated their considerable talents, it might have been possible to arrange a creative research link for DNA between King's and Cambridge. But in the end it was our MRC Directors, Randall and Bragg,[*] who had to decide policy, and their decision was that Francis and Jim should stop. Bragg was new to DNA, and I do not know what contact he had had with Randall over DNA policy (Bragg consistently gave me warm support) but the result was that a moratorium was set up. I continued to have friendly social contact with Francis and Jim, but we did not discuss DNA research. …

Back in the lab in January, I faced again the problem of DNA. Stokes had no brilliant thoughts, so I approached Hugh Longuet-Higgins, our new professor of theoretical physics. But he was full up with other scientific problems. I did feel very low. It was during these days that John Kendrew[†] rang me and asked if I would like to work with Francis and Jim. I said, 'What have they got to offer?' and he replied, 'They are very bright.' I thought he was right; but how could I avoid a big row over the Bragg Moratorium, and with Rosalind and Randall? I could not do it. …

Back in London I carefully mounted some of the fresh sperm on my high-resolution camera and got much clearer patterns than the previous year. The crystallinity was still less well defined than with good DNA patterns, and I wondered whether it was correct to use the term 'crystalline' when some irregularity was clearly visible. But when I met Bragg by chance I showed him the pattern and he reassured me. The pattern very clearly offered strong evidence for a helical structure for DNA. This was a very encouraging and

---

[*] Australian physicist Lawrence Bragg was head of the Cavendish Laboratory at Cambridge University.
[†] John Kendrew was using X-ray crystallography to study protein structures at the Cavendish Laboratory.

exciting result, and I wrote to Francis, including in my letter a sketch of the pattern—the striking X formed by the dots and shadows of the X-ray diffraction. …

For the first time we were emerging from a very dark winter.

The sharp sperm patterns were very inspiring, and had the special interest that sperm were real live objects and not just purified DNA extracted by chemists from living material. Much later I found out that Jim Watson had written to Delbrück* about my improved diffraction patterns of sperm, and had expressed his excitement with many exclamation marks. There were, however, complications in that there was protein as well as DNA in sperm heads, and in deriving the helical structure of DNA we would need to take into account the presence of the proteins. Francis had given me some advice on this. In parallel with the sharper sperm patterns, I had an encouraging thought about X-ray evidence that DNA was helical. Stokes had begun this before Rosalind had joined us. He noted that there was no diffraction along the length of the DNA apart from the clear 'pile of pennies' diffraction from the bases stacked on each other. That observation by Stokes had been the basis of all our helical thinking. The perfect continuity of the DNA helix showed itself by the absence of diffraction along the axis of the pattern. My new thought was similar. I knew that DNA molecules packed together in parallel like cylinders 20Å in diameter. I noted that in the outer part of the diffraction pattern (near the equator), there were regions where there was absence of diffraction. That fitted in with the DNA being helical and helped to cheer me up. But to be more certain, I needed better patterns.

Rosalind Franklin, however, soon announced the 'death of the helix', presenting Wilkins with new X-ray results that strongly contradicted a helical structure for DNA.

* Max Delbrück, professor of biology at Caltech and Watson's mentor.

Stokes and I went up to the pleasant airy room with two big windows that had been the head of department's before Randall moved down to our new basement building. Rosalind seemed in a quiet, sensible mood and told us about a recent collection of intensities from the crystalline A pattern. The directions of diffraction were at right angles to the fibre axis, or close to that. With a helical structure, the intensities would be the same for a reflection on one side of the pattern and on the opposite side. But Rosalind's careful measurements appeared to show clear differences between one side of the axis and the other. Stokes and I could see no way round the conclusion that Rosalind had reached after months of careful work. It seemed, in spite of all previous indications, that the DNA molecule was lopsided and not helical. I felt strongly that if DNA was not helical or some other simple form, it would be very difficult to find the structure. That would be very disappointing— the helix hypothesis had given us much hope and energy. ...

But the most disturbing question relates to an X-ray diffraction photograph that was, at this time, quite unknown to Stokes and me, and which was lying in a drawer in Rosalind's office when she made the 'death of the helix' announcement. This photograph—the now-famous 1952 pattern—was particularly clear, and everyone agreed subsequently that it provided important pro-helix evidence. I was given the pattern by Raymond on 30 January 1953, when Rosalind was preparing to leave our lab. Why would she, despite having found this evidence, give us, in July 1952, an account of why DNA was not helical? ... If she had shown Stokes and me the 1952 pattern, we almost certainly would have questioned her anti-helix stance.

Around this time, we all had to write a short account of our work for a report to our funders, the Medical Research Council. I expressed what I felt was a fair account of our work, and said that recent indications seemed not to point towards helices. I

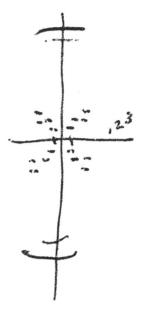

Above: Maurice Wilkins' X-ray sketch, from a letter to Francis Crick. *Below*: Rosalind Franklin's X-ray pattern, clearly showing the helix X shape. King's College, London

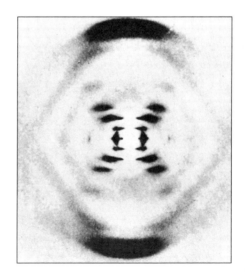

expressed some regret about that, saying that other more complex or less regular structures would be much harder to elucidate. The report was sent off to the MRC, and they circulated it with other labs' reports as part of their routine update to their scientists. To our colleagues, including those in Cambridge, it must have looked as though King's had given up on helices. …

Then, something extraordinary happened. One day in January 1953, Raymond met me in the corridor and handed me an excellent B pattern that Rosalind and he had taken. … It was much clearer and sharper than the first clear B pattern that Rosalind had shown us in October 1951—the one that had so excited Stokes and me. The new pattern showed the helix X shape more clearly than ever before. …

A few days later Jim was visiting us, and I stopped him in the main passage of our lab to show him the photograph. I said that it was very frustrating that Rosalind was continuing to base her work on non-helical ideas even though she had this new pattern that was even more convincingly helical than ever. As I stood with Jim in the corridor—I had the impression he was in a hurry to leave—I felt I must tell him what I had been thinking about base pairing in DNA. I had only recently had an idea and, because I respected him as a scientist and knew he had thought a great deal about DNA structure, I was eager to discuss the idea with him. In the event, however, I got no further than saying 'I think Chargaff's ratios are the key to DNA structure,' and Jim said, 'I do too' before he hurried off.

In our lab we had been well aware of Chargaff's ratios, but no one seemed to have realised that they might mean that the bases were paired in DNA. …

I came up with the idea of base pairing while trying to resolve the big contradiction in our DNA work: B-DNA seemed clearly helical whereas Rosalind's patient, thorough work seemed to show that crystalline A-DNA was not. To account for this contradiction, it occurred to me that DNA might be a lopsided helix with phosphates arranged regularly on the outside and bases lopsided inside—the big purine bases adenine and guanine on one side, and the small pyrimidines cytosine and thymine on the other. The bases would be hydrogen-bonded together in line with Chargaff's 1:1 ratios. Because I was dominated by the idea that there had to be three or more chains, my scheme had four chains with two base pairs lying beside each other. I was not very keen on the idea, because it seemed rather artificial. But it was an important step forward to think of base pairing at all. I noted that the distance across both pairs would be about the same, but I did not try to decide precisely where the hydrogen bonds were, or what the exact dimensions of the base pairs were. But my general idea of base pairs linking two chains in DNA was correct.

> Spurred on by chemist Linus Pauling's entry into the race to find the structure of DNA, Watson and Crick told Wilkins they had decided to get back to trying to build a model of the DNA structure.

Presumably Francis and Jim had spent some time considering their approach to me; but I was taken by surprise, and at first did not know what to say. ... But when I assessed the extent of the log-jam in our DNA work at King's, it seemed obvious that I could not ask Francis and Jim to hold off model-building any longer. And it turned out that Bragg was of the same mind: King's had had its chance—more than a year had passed and his moratorium should go. Also, it would have been too much for Bragg if Pauling were the first to find the DNA structure as well as the α-helix that he, Perutz and Kendrew had missed. It was, however, very disappointing for me: Rosalind was about to leave and, although there was some uncertainty about Rosalind's schedule which made forward-planning tricky, we did know she was leaving soon and I was looking forward to a positive, unified DNA effort in our lab. But we would instead feel a different tension—the London–Cambridge Rat Race was to begin again! What do you say if you are trudging towards the top of an unclimbed mountain and you see another group of climbers scrambling up a parallel path? When doing science, you can suggest working together—cooperation seems so much more creative than competition—but in the DNA case that did not seem possible. I do not remember thinking about the possibility that I might be The Big Shot Who Discovered the DNA Structure, but I did not enjoy making room for Francis and Jim. DNA was not private property: it was open to all to study peacefully without any one person throwing his weight about. I could see no alternative but to accept their position—I had principles, and science had to march on. But I was very cast down and could not conceal it. I had come up to Cambridge looking forward to a carefree jolly time, and now there was no chance of that. I just wanted to go home, and Francis had the sense not to press me to stay. As I walked out of the house, Jim came into the street and expressed his regrets; but I was not very receptive. I believe Jim meant well, but he later wrote, with admirable honesty, that irrespective of how I had answered their question, they would have gone ahead with their model-building. ...

When Francis and Jim had begun again it seemed likely that they would do much better than their first hurried model. I knew that they were thinking about base pairs because Jim had agreed with me that the Chargaff ratios were the key to the structure. I felt Francis and Jim were going to push ahead hard. There was no doubt we at King's must push hard too, and I had been looking forward very much to being with DNA again. To use Coleridge's expression, I loved DNA: I wanted to savour its nature and find what that nature revealed. The gardener turns the soil and reveals the insects, worms and other life which helps the garden to grow; a geologist studies rocks and the Earth's history is revealed. What would DNA reveal? Great prospects seemed to be emerging. There was a feeling of history in the air—and a pressure of necessity. I knew that our renewed attack on DNA would not be easy, but we just had to get on, keep cool and do our best. There were changes for the better. We could achieve unity in our work; no longer was I standing aside as Rosalind worked on and on to find non-helical DNA. She was leaving soon and had handed over to me a brilliant helical B pattern, and I looked forward to getting back some of the invaluable Signer DNA that she and Raymond had been using.

Until I knew definitely when Rosalind was going, I felt we could not set up a new group and get to grips with DNA. However, very soon after my ill-fated visit to Cambridge I remembered that I had arranged to visit Harriet Ephrussi in Paris about getting some of her biologically active DNA (it could genetically transform bacteria). … It was a really good little excursion and helped me to get going. We soon found Harriet's DNA gave a good crystalline X-ray pattern like that of Signer's DNA. This showed for the first time that the crystalline pattern, which gave the most detailed and exact X-ray data, was not just peculiar to DNA from a special type of cell in calf thymus—it was also given by real live genetic material. Thus, the wide significance of our work became even more clear than before. And, since Signer was no longer making his DNA, it was very good to have found an alternative source of first-class material— techniques for producing it had much improved in the three years since Signer had given away his DNA. And just a few days later, my friend Leonard Hamilton … who worked at the Sloan Kettering Cancer Institute in New York, sent us excellent quality human DNA from his lab. Leonard was to become our main supplier of DNA.

As Rosalind's departure became imminent, I began to assemble a group to continue, intensify and broaden our DNA structure work. The nature of the problem required a broad interdisciplinary approach. Also, like the Bomb Project, DNA structure was so important that a range of different approaches in parallel was needed. Randall agreed to it all. Bill Seeds was ready to give up microscope work and build molecular models of DNA. Now that we had really good DNA, Herbert Wilson put nucleo-proteins on one side and was full-time on DNA. Stokes looked forward to helping us with special mathematics. Now we could have a loosely organised group of the kind I had looked forward to when Rosalind had first arrived two years before. We could pull ourselves together, and it was very good to be working again in constructive comradeship.

I wrote to Francis on 7 March 1953, about a month after Rosalind's colloquium. … I had had a faint hope that giving him news of our broad attack on DNA might deter him from pushing on too fast.

It was John Kendrew, helpful as usual, who telephoned me to invite me to Cambridge to see the new model Francis and Jim had built, and he briefly told me what it was like. Jim later wrote that neither he nor Francis had wanted the task of breaking their news to me. This was not like the relatively carefree time 18 months before when Francis called me to see their first model. Now there was tension in the air. I was in a train to Cambridge straight away. And then: there was the model in front of me, standing high on a lab bench. In some basic ways the model was familiar: I recognised features as in the Fraser model—phosphates on the outside and bases stacked in the middle and joined by hydrogen bonds. But there were only two chains in the structure, and that made it very different from the Fraser model with its three chains. As I concentrated on Francis and Jim's model, Francis kept talking and confused me by referring to a diad axis, which made no sense because the bases were all different. Then I realised he was talking about everything other than the bases. He stressed that the special new features of the model lay in the hydrogen bonds that linked the bases in pairs as in the Chargaff ratios, and also in the important fact that the sequence of atoms in one chain ran up while in the other chain it ran down (that up and down was where the twofold diad axes came in). But the really impressive feature of the structure was the extraordinary way in which the two kinds of base pairs had exactly the same overall dimensions and shape. I had recently noted that the two Chargaff pairs would be about the same size, but I had not made detailed and exact studies. That exactness led Francis and Jim directly to a mechanism for gene replication. By splitting the base pairs apart, and pairing each base with another molecule of the same partner again, this DNA could replicate itself. Whether that idea was right or not, we could only guess. But a feeling came through to me that the model, though only bits of wire on a lab bench, had a special life of its own. It seemed like an incredible newborn baby that spoke for itself, saying, 'I don't care what you think—I know I am right.' (As the years went by, that baby had a lot more to say for itself.) Jim wrote later that sometimes he had feared that the DNA structure might turn out to be uninteresting. Clearly that was not the case. It seemed that non-living atoms and chemical bonds had come together to form life itself.

> Watson and Crick published an article on their ground-breaking discovery in the April 25, 1953 issue of *Nature*. In the same issue there were articles by Wilkins, Stokes and Wilson, and Franklin and Gosling, on the X-ray diffraction work supporting Watson and Crick's model.

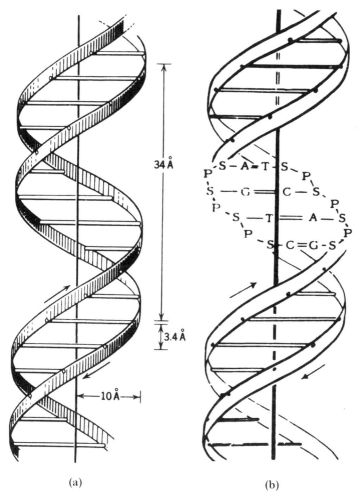

(a)   (b)

Diagram based on the figure in the original paper by Watson and
Crick showing the essential features and dimensions of the DNA
double helix—that is, the two sugar-phosphate chains running in
opposite directions, linked together through hydrogen bonded
base pairs stacked on top of each other. The separation of the
base pairs is 3.4 Å and there are 10 base pairs per helix pitch of
34 Å. In (b), A, C, G, T, S and P refer respectively to the bases
adenine, cytosine, guanine and thymine and to the sugar and
phosphate groups. King's College, London

In the years that followed, Wilkins and his lab continued their painstaking and exacting research, with their X-ray diffraction studies of the DNA molecule providing much evidence to verify the model that Watson and Crick had proposed. In 1962, Wilkins, Watson and Crick were jointly awarded the Nobel Prize in Physiology or Medicine. Rosalind Franklin, who had taken the now famous X-ray diffraction photograph that indicated a helical structure for DNA, had died in 1958, and so was not eligible for the prize.

From *The Third Man of the Double Helix: The Autobiography of Maurice Wilkins*: Oxford University Press, Oxford, 2003.

Making Waves:
A Poem for Maurice Wilkins

Making Waves by Wellington poet Chris Orsman was commissioned by the Royal Society of New Zealand in 2002
to celebrate the fiftieth anniversary of the discovery of the
structure of DNA. The celebrations also included the un-
veiling of a plaque on the site of Wilkins' former home in
Wellington, now part of Victoria University.

Light diffracted on a bedroom wall
at 30 Kelburn Parade, making waves
through a cloth blind, circa 1920;
outside, pongas and cabbage trees
lie just within memory's range,
a pattern and a shadow.
The silence here is qualified
but it draws you out, four years old,
or five. The world's a single room
where fronds and wind tap a code
against the window-pane.

Next up you're wild, sprinting down
a helix of concrete steps
from the hills to the harbour.
Or you're leaning into a gale
commensurate to your incline
and weight; the elements support you,
and the blustery horizon
is fresh with new information.

**

And now the landscape changes
from island to continent to island again,
and there's a sea-change as we fire off
certain rays to form a transverse
across your history.
                    Acclimatised,
you wintered over in laboratories
and made a virtue of basements
and arcane knowledge; you found
a scientific silence or a calm
in which things are worked out
at a snail's pace, a slime
stretched and scrutinised between
forefinger and thumb to yield
a feast of the truth, or a field
ploughed with frustration, if that
is where our guesses land us.
For Science is a railway carriage
rocking with big ideas, sometimes
stalled on the sidings or slowed
on branch lines near rural stations.
And still the whole is too huge for us
to comprehend, one metre long,
wrapped around each cell,
unread until it's unwound,
the scarf and valence of our complexity,
from which we derive our unique timbre
to say: *Well done! Well done!*

**

To an amateur an x-ray plate
looks like an old-fashioned

gramophone disk: yet it plays
scratchy music of the spheres,
jazz of an original order.
Or perhaps it's the ground-section
of a Byzantine cathedral, or a basilica
of double colonnades and semi-circular apse
– and who builds upwards from that
to discover the grand design? Who
constructs with only a floor plan
to find the elevations?
                              Those
who are neither architects nor masons
but quiet archaeologists of the unseen
hand and mind of God, digging upwards
to the exquisite airy construction
of the double helix. Gifted clumsiness?
Genius? You are there at the start of it,
a chiropractor of the biophysical,
clicking the backbone of DNA into place.

From *The Lakes of Mars* by Chris Orsman: Auckland University Press,
Auckland, 2008.

Establishment of a
Carbon-14 Laboratory

After the Second World War, Ernest Marsden, the head of New Zealand's Department of Scientific and Industrial Research—who, as a student, had worked with Ernest Rutherford on the gold foil experiment from which Rutherford determined the structure of the atom—established a nuclear sciences team. The staff included some of the young New Zealand physicists—including George Page and Gordon Fergusson—who had worked on the American and British wartime projects to develop nuclear bombs and nuclear energy. Marsden's initial plan was for New Zealand to construct a low-energy atomic pile to produce radioisotopes and act as the basis of an atomic research project. But after Marsden left the DSIR and the government failed to approve the plans for the atomic pile, members of the team began to move into other areas of research.

Athol Rafter (1913–96), a chemist who had joined the DSIR in 1940, spent 1948 with Gordon Fergusson, studying radiochemistry at the Massachusetts Institute of Technology and visiting nuclear establishments in the United States, Canada and England. At about the same time, the American chemist Willard Libby was pioneering the technique of radiocarbon dating, which dates organic material by comparing the ratio of radiocarbon atoms (carbon-14) to regular carbon atoms (carbon-12) with established radiocarbon to carbon ratios from the past.

On Rafter and Fergusson's return to New Zealand, they joined a re-focused DSIR nuclear sciences team.

Naturally, back in New Zealand, no one really knew what to do with us. Nuclear science was a very young baby clothed in mysticism and nuclear annihilation. Any expenditure on nuclear science was of necessity a major expenditure, and any work involved hazards that the people of New Zealand little understood.

We started off in the simplest way, George Page to build a mass spectrometer and Gordon Fergusson equipment to measure natural background radioactivity and to see what application it had in our thermal areas. I was given a cyclotron target and asked to separate radioactive iron, cobalt and nickel for use as tracers in agricultural experiments. Gordon worked in the attic of the then Dominion Physical Laboratory, George Page in a relatively good-looking outer shed and myself in a shed that looked like an outhouse.

In March 1950, Dr Falla of the Dominion Museum contacted one of our assistant secretaries, Charles Watson-Munro, expressing interest in the determination of age by radioactive carbon methods. This was in itself rather unusual, for the paper on which Dr Libby was then the most junior author, 'The Half-life of Radio-carbon (C^{14})' that appeared in *Physical Review*, June 15, 1949, did not reach New Zealand until April 5, 1950. I eventually received a request to look into this new radioactive carbon method.

A few days later I was walking home quietly through the grounds of our Parliament Buildings when coming in the opposite direction was the head of our department, Mr Callaghan, who stopped me with the statement, 'Rafter, I have just come from a meeting of geologists who tell me that there is a method of dating by means of carbon that should be able to tell the age of our volcanic ash showers. Would you see if you could develop this method and stop the geologists arguing.' I said a confused good-night and continued on my way home somewhat more puzzled than usual.

I read the review article and was not very much wiser, so I wrote to a friend I had met while visiting Berkeley in 1949, Dr Andrew Benson, asking him if he knew this fellow Libby and

251

saying that the details I could get from the paper were very scant. Dr Benson replied that he knew Dr Libby, but from what he knew the methods were changing too frequently to justify detailed publication. He suggested I should write to Dr Libby.

I had enough courage to write to Dr Libby on May 22, 1950, before which time his article in *Science* had appeared on 'Age Determination by Radiocarbon Content'. My problem was to get carbon on to the inside of a screen wall counter. Dr Libby kindly took pity on my request and asked Dr Jim Arnold to reply. He kindly and clearly explained how they spun the carbon on to the inside of a copper cylinder. Now for those of you who never attempted the solid carbon method of $^{14}C^*$ dating I feel you have missed one of life's great experiences. If I was asked for my opinion of Dr Libby, I would say he was the proverbial Job of science, for I wonder how many hours he spent trying to get carbon to stick on to the inside of a copper cylinder. But I had maybe even a more difficult problem. Once I had the carbon sitting nicely around the walls of the cylinder I now had to get it into the counter. My little laboratory happened to be 12 miles from where Gordon Fergusson was with his counter in the attic waiting for it. Now, that 12 miles was not along a smooth American superhighway but a frequently much potholed road, particularly near the Physical Laboratory, where all too frequently my sample collapsed and we returned home again. However, mainly through the ingenuity of Gordon Fergusson, in July 1953 we actually did publish the first list of New Zealand ^{14}C age measurements of six results, four dating the Taupo ash showers that it had been initially our objective to achieve.

However, I think rather fortunately, I ran into trouble again. The problem with the carbon method was the difficulty of

---

* Carbon-14, or radiocarbon.

removing contaminants that were in the magnesium used for reduction of the carbon dioxide. I was unable to obtain supplies of magnesium from America, and when I finally found an English supplier I ordered several kilograms of the metal, only to find that it was so contaminated with radiochemical impurities that finally, in October 1953, we gave up the solid carbon method altogether. Gordon and I decided we should be able to make carbon dioxide into a better counting gas. Gordon soon had the electronic equipment running for counting carbon dioxide in a proportional counter but what annoyed me was that while he could get CO_2 straight out of the fire extinguisher to count, anything I at first produced was rather useless. However, I was convinced that if it could be commercially made pure enough then chemically it also should be possible. This was eventually achieved by early 1954, just in time for the enormous potential for the radiocarbon method that lay ahead.

The great advantage of the CO_2 proportional counting method was the much greater accuracy possible. Having achieved this favourable position, it was another thing to do something useful with it. Page had built a mass spectrometer, and as the chemist responsible for putting something into it a decision had to be made. Here again we were fortunate, for many years ago my early training had been as a coal chemist and then a rock analyst and I knew that our New Zealand coals had in them as much as 10% sulphur. I thought it would be a good project to attempt to solve the origin of this sulphur by the then beginning of the science of biogeochemistry. We had to get SO_2 into a mass spectrometer. As we could now measure ^{14}C variation with an accuracy of $\pm 0.5\%$, it looked an excellent opportunity to study the ^{14}C variations in nature. This led to our work on the atmosphere and coastal water in Makara, Wellington, the study of the ageing of the New Zealand moa, the survey of the surface ocean waters and problems with

Antarctic specimens, and other aspects of the ever expanding possibilities of the radiocarbon atoms.

But in many of these developments the work at our institute would have been rather worthless without the guiding hand of mass spectrometry that, to our carbon-14 laboratory, is of fundamental necessity. Now, I could spend more time than you would care to waste discussing the problems of the carbon-14 laboratory but the guiding principle must be 'you must never give up'. The CO_2 method is a most beautiful method, and, like all things beautiful, has a most cantankerous side, the presence of the electronegative impurities that drive technicians to despair, scientists to drink and harmony to discord. But like the suffering of pain, the relief is so much greater when all is well again.

> In 1959, Rafter became the inaugural director of the DSIR's Institute of Nuclear Sciences, a position he held until his retirement in 1978. Athol Rafter is now considered one of the pioneers of radiocarbon dating, with among his most significant contributions his discovery, with Gordon Fergusson, of what they called 'The Atom Bomb Effect'— rising levels of radiocarbon in the atmosphere caused by atmospheric nuclear weapon tests.

Adapted from 'Problems in the Establishment of a Carbon-14 and Tritium Laboratory' by Athol Rafter, a paper presented at the Sixth International Conference on Radiocarbon and Tritium Dating, held at Washington State University in 1965.

The Allende Meteorite

In 1939, after graduating from the University of Canterbury, Brian Mason (born 1917) travelled to Norway to study for a PhD in geochemistry. His plans were disrupted by the German invasion of Norway and he escaped to Sweden, where he completed his thesis. Mason then moved to the United States, where he had a long and distinguished career as a geochemist specialising in meteorites. In 1965 he took up a position in the Division of Meteorites at the Smithsonian Institution in Washington, DC. The American space programme was in full swing and it was an exciting time to be involved with extraterrestrial rocks. The 'chondrites' to which Mason refers throughout this piece are stony meteorites formed from the accretion of dust and small grains present from the birth of the solar system.

The year 1969 was the *anno mirabilis* for researchers on extra-terrestrial material. On Saturday, February 8 at 1.05 a.m., an exploding fireball scattered tons of the Allende meteorite over the countryside near Parral, Mexico. In July, the Apollo II astronauts made the first landing on the moon and brought back 22 kilo-grams of lunar material. On September 28, a shower of meteorites near Murchison, in Victoria, Australia, provided more than 100 kilograms of a carbonaceous chondrite. And in December a team of Japanese glaciologists picked up nine meteorites on the ice cap near the Yamato Mountains in the African sector of Antarctica. It was not realised at the time, but this was the first intimation of Antarctica as a rich collecting area for meteorites.

I learned of the Allende fall through a brief note in *The New York Times* for Sunday, February 9, 1969, describing a brilliant fire-ball seen in Texas and northern Mexico. When I got to the museum the next day we had a consultation and decided that this event deserved investigation on the ground. However, it would not be politic to venture forth without the approval of the Mexican authorities. I telephoned the director of the Instituto de Geologia, Diego Cordoba, for permission to investigate the fall area and collect material for scientific investigation. He agreed, and the following day Roy Clarke and I flew to El Paso and rented a car. We drove into Parral on Wednesday afternoon (February 12). Roy was driving, and as we passed along the main street I suddenly saw in a newspaper window a large black stone that looked like a meteorite. We made a quick stop, and I dashed out to inspect it. It took only an instant to recognise it as a carbonaceous chondrite, and on a broken surface I saw large white inclusions, the like of which I had never seen before. I was jubilant—we had hit pay dirt!

We booked in at the local hotel and contacted Mr R.C. Byrd, superintendent of Asarco Mexicana, which had a large operation in the Parral area. He was most helpful and lent us a Mexican engineer, Manuel Gomez, who was invaluable as a liaison with the local people. The following morning we drove to Pueblito de Allende, a small village where stones had fallen. Discussing our plan of operation, Manuel suggested we recruit a search party from the local school. The headmaster was most cooperative; he assigned us his sixth-grade class of thirty-eight boys, who were enthusiastic for a day in the field. We rented a truck, bought a couple of cases of Coke for refreshment, and set forth. We posted our recruits in a long line and directed them towards the next village, about three miles away. Within minutes we had our first meteorite, a beautiful crusted stone weighing about five kilograms, sitting on dry grass (the meteorite could not have been hot when it landed, since the grass was not even scorched).

We spent five days in the area, certainly the most enjoyable and fruitful collection trip I have experienced. It was early spring, the weather sunny and mild, and the fruit trees just coming into bloom. The terrain was ideal for searching—dry plains country, partly ploughed for spring planting, and terrestrial rocks mostly of limestone, not readily confused with meteorites. We collected and acquired by purchase about 150 kilograms of this rare carbonaceous chondrite (more material than all the previously known carbonaceous chondrites). With many laboratories geared up to analyse lunar rocks, this was indeed manna from heaven. In a few weeks we supplied Allende material to scientists in thirty-seven laboratories in thirteen countries.

It is interesting to consider the fortuitous coincidence of the Allende fall with the imminent return of the first lunar rocks. This coincidence clearly stimulated the intensive research devoted to Allende. When I looked in our collection I found there were similar meteorites which had fallen or been found many years earlier—Kaba, Bali, Leoville for example. But these were small meteorites, inadequately described, and had attracted little attention.

Allende was full of surprises. Within a few weeks I had identified minerals previously unrecorded in meteorites—grossular, andradite, sodalite, gehlenite, and a unique Al/Ti-rich pyroxene (up to 22% Al_2O_3, 18% TiO_2). Just after we returned to Washington with our booty, I was visited by an old friend, Louis Fuchs of the Argonne National Laboratory. Louis was an accomplished mineralogist with a remarkable eye for the interesting and unusual, and I let him scan through our collection and select material for his research. In one piece he selected he identified cordierite, previously unknown in meteorites. Neither I nor any other mineralogist has found another specimen with cordierite, despite intensive search!

The recovery of a large amount of Allende enabled the preparation of a meteorite reference sample for chemical analyses. In the Division of Meteorites we had discussed this in the past. We decided that it should be a chondrite but the difficulty of producing a homogenous powder of an ordinary chondrite containing 5–15% of nickel-iron could not be overcome. Allende provided the solution. The meteorite was a fresh fall available in large quantity. It was a carbonaceous chondrite containing only trace amounts of metal in microscopic grains, hence easy to crush and homogenise. We took about four kilograms from a 35-kilogram specimen … for this purpose. During the crushing I extracted numerous Ca/Al-rich inclusions, which provided me (and others) with a remarkable and unique variety of research material. The reference sample was analysed by 24 geochemists for a total of 74 elements. The results were compiled and published and splits of the reference sample have been used as a standard by laboratories throughout the world.

Much of my subsequent meteorite research was on the composition of these Ca/Al-rich inclusions. Mineralogy and bulk compositions were determined in Washington, minor and trace elements on the spark-source mass spectrometer in Ross Taylor's laboratory at the Australian National University in Canberra. The biggest surprise was the remarkable fractionations of rare-earth elements. Ross and I eventually distinguished six distinct distribution patterns for these elements, and remarkable anomalies in the abundances of europium, thulium and ytterbium. We also analysed two splits of the Allende reference sample for 22 trace elements, and our results were in excellent agreement with the eventual recommended values.

The Allende meteorite has been truly called 'Cosmochemistry's Rosetta Stone'. As one of my colleagues remarked, 'We have learned more about the origin and evolution of the solar system from the Allende meteorite than from the lunar rocks—and it came to us free.'

From *From Mountains to Meteorites* by Brian Mason and Simon Nathan: Geological Society of New Zealand, Lower Hutt, 2001.

Premature Birth
and Respiratory Distress

Graham (Mont) Liggins (born 1926) followed his father's
career footsteps into medicine. In the 1940s he trained as
a doctor at the University of Otago, and then travelled to
the United Kingdom to specialise in obstetrics and gynaecology. On returning to New Zealand, he branched out
from clinical practice to research the timing of birth, and
made a discovery that dramatically improved the survival
rates of prematurely born babies.

I came back from the United Kingdom in 1959 and was appointed to an academic position in obstetrics/gynaecology at
the University of Auckland. I looked for a research topic to pursue.
My friend and colleague Bill Liley, of foetal transfusion fame,
suggested that I find something that was important and potentially
soluble. At that time much the biggest killer of newborn babies
was respiratory distress, which is the result of immature lungs. So
clearly premature delivery, which was the cause of respiratory
distress syndrome, was the important topic. However, it seemed
to me that if one were going to work out the cause of premature
labour, one had to work out first of all why labour occurred
normally at term.

I decided it was worth revisiting Hippocrates' idea that it's
the baby and not the mother who determines the time of birth.
There was circumstantial evidence pointing to a role for the
foetal pituitary in initiating labour and I decided to test this idea
experimentally. I worked out a technique for destroying the pituitary in foetal sheep. This was before anyone had found a way
of getting access to the foetus, operating on it and returning it

to the uterus and having the pregnancy continue. So, spending my weekends working on sheep at the Ruakura Research Station in Hamilton, I developed a technique that was fairly simple if you knew how. You picked up the foetal head—which was still inside the uterus—in your hands, then passed a fine probe down along the floor of the skull until it hit the back of the pituitary fossa. You then applied the cautery and that destroyed the pituitary.

I then spent a sabbatical leave in the Veterinary School at the University of California, Davis. There I collaborated with the pathologist Dr Peter Kennedy and found that sheep whose foetuses had had their pituitary glands removed failed to start labour. I recall the mounting excitement as the first sheep reached term and then continued day by day until interrupted by caesarean section. This really sent ripples around the world. There had been no progress for many years in the field of labour and laboratories around the world immediately took it up.

I came back to Auckland to find no facilities and no money, but was able to get research funds from the Wellcome Trust in the United Kingdom. I bred my sheep on the Auckland showgrounds, across the road from the sheds we were using as a laboratory on the grounds of National Women's Hospital. When I'd finished with my sheep and they'd been put down, it was then a matter of disposing of their bodies. For a long time I used a facility at the old National Women's Hospital in Cornwall Park. I'd finish up my day's work and on my way home put a carcass down a chute into the incinerator. I had an old Mercedes that I'd shipped back from Britain. My kids were always complaining about the smell of sheep. One morning I had to admit that it was a bit overwhelming. I opened the boot and there was a sheep I'd put in several days before and forgotten about. Dead sheep swell up like crazy, like balloons, from the gases in their rumen.

As soon as I got the opportunity I went over to the incinerator and, with a lot of difficulty, managed to get the sheep down the chute—it was a very tight fit. I pushed it down, turned my back, and was walking back to my car when suddenly there was a huge explosion. This sheep had just blown up! A shower of ashes went up the chimney. There was an ash-catcher over the chimney, but both the ash and the ash-catcher blew a couple of hundred feet up in the air. By the time I got back to hospital the town clerk had already been on the telephone to the hospital. Apparently his wife lived in Wheturangi Road, and all the ash had blown on to the washing that she'd just hung out. So the incinerator was closed down; we then had a lot of difficulty finding a suitable way of disposing of the sheep.

The next step was to find out how the action of the pituitary was mediated. The adrenal gland seemed a likely candidate. Foetal adrenalectomy confirmed it. We then tried to find which hormone was responsible. Cortisol—a steroid hormone produced by the adrenal gland—was tested by infusing it into the foetus, and sure enough, it induced premature labour. It seemed that we were well on the way to solving preterm labour. (Unfortunately, although subsequent work confirmed the role of cortisol in many animal species, it was not so for humans and other primate species.)

At the point where we were infusing cortisol into the foetuses, serendipity stepped in. At the time when our premature lambs were arriving they had no right to survive. They should have died immediately after birth, because their lungs were too immature for them to survive. But to our great surprise we found premature lambs that had been born during the night were breathing in the morning. And at autopsy we found they had partially inflated lungs, which they had no right to do. Clearly, cortisol was inducing changes in the lungs that accelerated their maturation. We guessed, and it turned out very quickly that the guess was right: cortisol was inducing the surfactant material in the lungs that is responsible for keeping them inflated and stable, allowing them to resist the forces of surface tension that tend to lead to collapse of the lungs.

The exciting possibility arose that the effect of steroids on foetal sheep lungs might apply also to human foetal lungs. I was aware that steroids given to pregnant women crossed the placenta and depressed foetal adrenal function, so the stage was set for a clinical trial in human pregnancy. In collaboration with my neonatology colleague Dr Ross Howie, a double-blind, controlled trial was designed and carried out in a large number of women in threatened preterm labour. Mothers at risk of early delivery were randomly assigned to either standard treatment or standard treatment plus two doses of steroid. Two hundred and sixty-eight women participated and it was a watershed study.

The results, published in 1972, showed a dramatic reduction in respiratory distress syndrome and neonatal mortality in treated infants. Mortality was halved. This included babies born three months or more before term. Nine controlled trials in various parts of the world confirmed our results.

Considering that this was one of the very few perinatal treatments shown to be effective in a controlled trial, its adoption in practice was remarkably slow. This was attributable mainly to unwillingness of professional bodies to support the treatment. It was nearly twenty years before the United States National Institutes of Health and the British Royal College of Obstetricians and Gynaecologists made firm recommendations in its favour. Initially, too, there were concerns about long-term adverse effects and neonatologists tended to resent what they perceived as obstetricians intruding on their territory.

Liggins' technique of administering steroids to pregnant women experiencing premature labour is now standard international practice and has saved hundreds of thousands of lives worldwide. Liggins was made a Fellow of the Royal Society of London in 1980 and knighted in 1991. He is now Emeritus Professor of Obstetrics and Gynaecology at the University of Auckland, with the distinction of having a biomedical and clinical research body—the University of Auckland's Liggins Institute—named in his honour.

Adapted from *A Brief Autobiography of Graham Liggins FRS* (unpublished); and an interview by Allan Coukell with Graham Liggins on *Eureka*, Radio New Zealand National, 3 February 2002.

Valley of the Dragons

The discovery of New Zealand's first dinosaur fossil—evidence that dinosaurs once walked New Zealand's forests—was made by an amateur palaeontologist, Joan Wiffen (born 1922). This Hawke's Bay mother and housewife became interested in fossils in the 1960s after taking night classes in geology. In 1972, inspired by a map that described 'reptilian bones in beds of brackish water in the Te Hoe Valley', Wiffen and her family 'loaded the car with lots of food and drink, a billy to boil water in if required, road maps and our hand-drawn map for beyond the fringes of civilisation' and set off in search of dinosaur bones. After hours of driving, following their map down gravel roads through farmland, they parked beside a bridge running over a stream in Hawke's Bay hill country.

So, it was out of the car and a hurried clamber down the steep bank, and there they were—fossils in profusion. Every one of the cold grey stones in the water seemed to sprout fossils, while green moss-clad rocks littered the verge and ferns and shrubs covered the banks. True, there were no dinosaur skeletons lying around in the boulders or in the crumbling shale bank where the bridge was bedded, or any bones at all for that matter, but all this was forgotten in the excitement at what we saw. There were rocks encrusted with fish teeth, shark teeth, fish scales and vertebrae, gleaming on the surface where the rock had been worn away in the swift-flowing water, leaving these harder, dark, shining remains etched out. There were shells of all kinds, a few worn belemnites and a fragment of a small ammonite, a straight form which I'd never seen before, but unmistakable with its typical suture lines. Time passed in a flash, and soon we realised the sun was disappearing behind the surrounding hills and the light fading as we came out of the creek, cold, wet and bemused.

The family returned again and again to the Mangahouanga Stream during the summer of 1973.

The stream was difficult to negotiate, with its cold fast-flowing water, slippery moss-covered boulders and steep banks, so initially we didn't move far from the bridge. Since there were so many fossils lying around, there was little need. Later, when we ventured further afield, it was difficult to find new points of entry to the stream bed—the banks were slippery underfoot and rough with undergrowth. Frequently, when we found a way down, we were confronted with a considerable drop down the sharply undercut bank to the water.

At this stage we used the stream bed to explore the area as far as possible, but big, deep, black pools, waterfalls, rapids and even log-jams often made this impossible. It then became necessary to clamber out, before dropping back into the creek again at the first opportunity. Later, we made small tagged walking tracks to the creek from the nearest road access. Collecting up- and downstream from these entry points, we could cover six to eight kilometres of stream bed and reduce the distance over which we carried our equipment. In addition to suitable clothing, packs and tools, we now carried a rope, which was often secured to trees along the narrow tracks as a safety measure. It also helped us haul our weary bodies and heavily laden packs up out of the creek bed at the end of the day.

Buoyed on by the success of some early fossil finds, mostly shells common to marine sediments, Joan and her husband Pont upgraded their calico tent to an iron-roofed wooden hut and continued their search.

It was later in 1973, when we had learned to negotiate the under-growth, the lawyer vines, the cutty grass and the slippery narrow tracks down the banks, that we discovered a new, narrow, fast-flowing section where we were able to cross the stream. From this point we made our way cautiously upstream and, rounding a curve where the shallow water tumbled rapidly over a stretch of dark, moss-wet rocks, Pont bent down and shouted … there was our first fossil bone!

It was a moment of intense excitement as we gathered around and bent down to look in wonder at the black waterworn bones exposed on the surface of this grey sandstone boulder. The rock was dislodged from the bed, turned over, lifted and carried carefully to the bank where we looked reverently at the bones of some animal that had lived 70 million years ago. These were undoubtedly the saurian bones mentioned on the map that had led us to this site. As I ran my hands over those ancient, cold, wet bones, I felt a surge of excitement—this was our initiation into vertebrate palaeontology, the history of early New Zealand life.

> The sediments Joan and Pont Wiffen were exploring were 65 to 70 million years old, once part of the eastern coastline of New Zealand, where 'the long-necked plesiosaurs and the large, predatory, alligator-like mosasaurs lived, bred and probably raised their young, while they fed on the abundant varieties of shark and fish'.

Those first fossil bones that Pont found in 1973 were later identified as plesiosaur vertebrae, and plesiosaur remains proved to be the most common bone fossils found at the Mangahouanga site. Of these, the discovery of a complete skull in 1978 was by far the most exciting. It was the first found in New Zealand and one of less than a dozen complete elasmosaur skulls known anywhere. The recovery of this specimen resulted from an early experiment

Above: The long-necked plesiosaur. *Below*: The mosasaur, a fierce marine predator. Illustration by Geoffrey Cox

with explosives—not a desirable method of breaking fossiliferous rocks, but useful for large boulders in water which can't be moved in any other way. The disadvantage of using explosives is that it tends to shatter the rock, and the bones, being the weakest structural part of the rock, suffer most. This skull came out in around eight or nine large pieces of rock and a number of small fragments, all of which were carefully wrapped, packed and carried out, then stored until we felt capable of the delicate task of total extraction, in 1984.

> The seas of ancient New Zealand were also home to another large reptilian predator—the mosasaur.

The first major mosasaur specimen was collected from the site in 1974, an episode we all remember very well. We had seen a large boulder in deep water well upstream, with a number of bones exposed on the surface below water. The rock was too big to move, despite our efforts with crowbars and ropes, so it was at last decided to try explosives to move and perhaps crack the boulder, in order to get it into a workable position.

This was done, with shattering results. We surveyed the pieces with both dismay and delight. Though large jagged chunks of rock still confronted us, we saw a massive amount of bones. We then prepared to remove the pieces from the stream bed and convey them back. With a coil of No. 8 fencing wire, we erected a kind of flying fox—the wire was tied between suitable trees along the route, finishing well up the ridge overlooking the fossil site. Then we all got to work, starting at the creek with bags and sacks, which we wrapped around the rough-edged rocks after getting them out of the water on to the narrow ledge alongside. From here we attached the individual bundles to a rope and, with someone heaving from below and someone hauling from above, we finally got all thirty pieces of rock to the top. They were then tied to the wire, and then they were slowly slid, edged, dragged and hauled along, finally arriving at a point above the road. Here they were rolled down the bank to the car and, eventually, taken triumphantly home. It was found later that some parts seen on the day—bones and particularly some teeth—were missing. It was not until the extraction work was underway that we returned in better conditions, retrieving the missing portions of the boulder, and the precious mosasaur jaw still lodged in the creek bed sediment at the same place!

This specimen was unloaded at home, numbered and stored until extraction commenced in 1978. In 1980 it was described before being given to the National Collections. Two more mosasaur skulls were collected in 1980 and 1982, and later extracted and described, along with other interesting mosasaur material. As a result several distinct types of these rapacious reptiles have been identified from this site.

> Charles Fleming had in 1967 suggested that the lack of a dinosaur fossil record in New Zealand might be simply because no fossils had yet been found, but the general consensus in the 1970s was that the New Zealand land mass had never been home to dinosaurs. Dinosaur fossils eluded Wiffen and her team until 1974.

We had found an abundance of marine fossils, all well on the way to documentation, but there still remained the mystery of land fossils, or rather the lack of them—we had not identified any so far. ...

That New Zealand had a forest back in Jurassic times (more than 135 million years ago) is well-recorded, but did these ancient forests resound with animal- and bird-calls, or were they silent and the land without life? Again we have to rely on the fossil record to find out, and the land sediments from that period have proved unfossiliferous. The high rainfall and rapidly eroding forested landscape produced acidic conditions that were un-favourable to the preservation of bones or shells, so that there is virtually no record of terrestrial life from that time. The failure to discover any dinosaur bones had generally been accepted as evidence that they had not spread across this part of Gondwana, before New Zealand's separation. ...

Scientific logic ... suggests that to find land dinosaurs one should look at land sediments, a fact of geological life that we, as amateurs, had overlooked when we originally set out to find a dinosaur at our marine site. In fact, the question of what sort of rocks we were looking at had been of little concern—we knew their age and that they contained fossil bones, and that was enough for us. ...

Meanwhile, the abundance of fossils already found at the site provided us with enough excitement and incentive to continue our exploration of this area for whatever kind of fossil turned up. So we went on collecting as time and weather permitted and, with the experience gained from the now massive range of marine reptile, fish and shark remains, it was possible for us to recognise the dif-ference between these bones and the other odd ones that turned up. Immediate recognition of fossil bone in the rock is rarely possible. By this stage we usually knew if it was plesiosaur, mosa-saur, fish or turtle we had found, from the shape or structure of the exposed bone surfaces seen in the rock. Gradually the collection of unknown bones and bone fragments grew, but without the means of ready identification the first excitement of discovery dimmed a little and they were stored or put away in a cabinet, to be brought out to show any visitors, in the hope that someone might recognise a specimen.

The first real excitement came unexpectedly in 1974, when Bill brought me a piece of rock and asked me to try my recently acquired skill of acid extraction. In an attempt to reduce the size of the rock, he broke it in two, revealing a very large fish vertebra in addition to another fossil he thought was an ammonite. While preparing it, I saw a small hollow brown bone, broken in two, exposed on the rock surfaces. I carefully coated these fragments with the protective resin, and placed them in the prepared acid bath. Slowly, after repeated washing, cleaning and recoating, the fossils emerged.

My interest was centred on the small hollow bone, because it was unusual, different from all the other material in the rock. After a couple of weeks, the job was done. Bill's ammonite proved to be a *Chimaera* beak; the large fish vertebra was glued back together again, and the two pieces of the small, hollow brown bone were removed, washed and carefully rejoined. We had a perfect toe bone. Judging by the shape of the two articular ends, this was certainly not a toe bone from a marine reptile, but from a land-walking animal. My only knowledge of such a bone was the foot of a moa, and this toe bone seemed to have the characteristics of those seen in that bird's foot. This caused great excitement in the household, but where to from there?

I consulted Canterbury Museum's osteologist, Ron Scarlett—a very experienced bird man—and he too thought that this bone was probably from an early flightless bird, more likely the extinct rail than a moa. Then another opinion cropped up, unexpectedly. Dr Dale Russell, from the Museum of Natural History in Ottawa, Canada, and his team were on a world tour testing for iridium[*] on the Cretaceous–Tertiary boundaries, and they called in to look at

[*] Iridium is a silvery-white chemical element that is geologically distinctive for its presence in a thin layer of clay that marks the boundary between rocks of the Cretaceous and Tertiary eras, about 65 million years ago. It is believed that this 'iridium layer' is of extraterrestrial origin, and came from a meteor that collided with the Earth, leading to the extinction of the dinosaurs.

our fossils. Dale was an expert on mosasaurs as well, so his views on our reptile bones were eagerly sought. While looking at our fossil collection he also noticed the small toe bone, and said, 'Oh, I see you have a theropod toe bone.'

I searched my mind feverishly to recall what a theropod was, and told him that we had thought the bone was from a bird. This raised doubts all round—bird or theropod dinosaur? As far as we were concerned it didn't matter too much—we had a land fossil, a toe bone from a bird-like foot. However, writing a scientific paper to describe it was delayed, and it was put aside until more material or information could be found. …

Sometime after this, in 1975, Pont and I returned to the fossil site, and found a new bone. It was in a football-sized concretion, broken on the site to investigate the wood showing on the outside. Inside, lying near the pieces of fossil wood, was a bone, now broken into several pieces. I picked the fragments up, and looked at them curiously. It was a rather unusual sort of vertebra, so the pieces were wrapped carefully and taken home. Later they were fully extracted and put together again. The bone was obviously not from one of our marine reptiles and I couldn't find anything similar in the textbooks I regularly used. The new specimen obviously came from the backbone of a different animal, but information on vertebrate fossils was hard to come by, and at that time I wasn't sure what other reptiles we might be expected to find at the site. So, listed and numbered in my record file, it joined other unidentified fossils on the shelf in the cabinet. …

Our collection of unknown bone fossils and fragments grew slowly. All were extracted, numbered and recorded on a fossil file and set aside to be taken out and examined from time to time, or shown to visitors in the hope that someone might turn up who would recognise them from their own personal experience.

In 1979 Pont and I went to Brisbane for a holiday, and I wrote ahead to the museum and asked if we could look at their vertebrate fossils while we were there. They had a huge plesiosaur, called *Kronosaurus*, and lots of other strange marsupial fossils we looked forward to seeing. We were most kindly received by Dr Mary Wade and her staff, and were introduced to the recently appointed curator, Dr Ralph Molnar, an American palaeontologist. This was our first encounter with a real live dinosaur enthusiast.

I had taken a few small casts and some colour slides of our site and work, in which the staff took an interest, and we, in turn, were amazed to see the work that was being carried out on the preparation of some Queensland dinosaurs. It was while talking to Ralph, in his office, that I noticed a couple of vertebrae on his desk—and I was riveted. I looked at them disbelievingly and said, 'I have a vertebra like that at home. What is it?'

'A land dinosaur … an ankylosaurus,' he said, picking up the fossil and handling it reverently. 'They are often found in or near water. Are you sure it's like this one?' he asked. He suggested that I send a cast or photograph with measurements to him when I went home, and he would be able to tell us if ours was a dinosaur bone or not. The rest of the holiday dragged for me. I couldn't get home quickly enough. The cast of the specimen was made and sent with photographs, and any other information I could find, and about a week later, I had a phone call. 'It's a dinosaur!'

My excitement knew no bounds—at last we had a dinosaur, not an ankylosaurus, but most probably a theropod. It didn't matter too much to us what family it belonged to, we weren't into family trees—we had evidence of a Mesozoic dinosaur here in New Zealand at last.

Wiffen then sent the original bone to Ralph Molnar, who studied it in preparation for a scientific paper describing the fossil find.

It was decided that it would be appropriate for Dr Molnar to read the paper describing our dinosaur vertebra at the fifth Gondwana Symposium, held under the auspices of the Royal Society of New Zealand in Wellington, in February 1980. There was no question of bypassing a New Zealand expert—there were none.

Because this was to be the official recognition of our dinosaur, we had informed the Geological Survey, with whom we worked closely, but we made no public announcement of the discovery before the conference.

At last the day came. I went to Wellington's Victoria University, where the symposium was being held, met Ralph, received the vertebra back and handed it over to the staff of the Geological Survey for safe keeping in the National Collections. A small problem arose when the editors of the ensuing publication of papers from the symposium asked for the place of repository and registration number of the fossil. The repository was the Geological Survey, but when the survey were asked for a registration number they said they didn't have one. There was no register for fossil vertebrates. The reply was, if you want this bone, you have to produce a registration number for it—and that was the origin of a new New Zealand fossil file register, C.D. (Chordata, backboned animals). Number 1 was our dinosaur tail bone.

Dr Molnar's paper was scheduled for Wednesday afternoon and I, along with a number of other people from New Zealand and overseas, settled in the lecture rooms. I sat there in a state of nervous expectation, with all the maternal pride of a mother at the christening of her first child, while the others remained speculative, mildly interested or somewhat disbelieving. Ralph had done considerable research to try and establish a creditable ancestry for our beast, with some difficulty. Although it was unquestionably a dinosaur tail bone, it was not entirely like any others previously described. However, this was reasonable, considering the isolation and the period in which it was preserved. Now at least it was described in detail and recorded for posterity as New Zealand's first dinosaur—a theropod, a carnivorous bipedal dinosaur of moderate proportions.

The paper was delivered, there were a few polite formal questions, congratulations to Ralph on his paper, and a handful of people I knew from the Geological Survey came to me to express their interest and congratulations on the dinosaur discovery. Later in the day, on my way back to the university hostel, a neat gentleman, with a bow-tie, came up to me and said, 'I'm Charles Fleming, and I'm delighted to hear about your dinosaur.' And so I met Sir Charles Fleming, the man who, many years before, had recognised the possibility that dinosaurs existed in New Zealand. It was the most exciting moment of the day for me, that simple acceptance by one of New Zealand's great scientists.

But, for the greater part, the reaction was a thunderous silence, and a general lack of interest or understanding of the geological significance of dinosaurs in New Zealand. What had I expected— a champagne party? Looking back, the disappointment I felt that so few shared in our excitement over the proof that New Zealand had dinosaurs was somewhat naïve. The typically con-servative reaction was perhaps to be expected. After all, it was only a single vertebra, located at some obscure place that wasn't even a recognised fossil area, by a bunch of amateurs. Nothing to get carried away over, probably washed over from Australia anyway. The competent, conservative and, in some ways, very complacent world of geology had succeeded quite satisfactorily in New Zealand without the intrusion of dinosaurs. The study of invertebrates for geological purposes had served well without the need to recognise terrestrial vertebrate fossils and their geo-logical significance in relation to the age and origins of this land,

Carnivorous theropods of the kind known to have lived
in New Zealand. **Illustration by Geoffrey Cox**

so perhaps the reluctance to accept any change on such scanty evidence was understandable. …

Undiscouraged, we continued our search for fossils. Of course, we hoped for more dinosaurs—large skeletons or great skulls with gleaming teeth, such as we had already found with the marine reptiles. But fossil collection rarely happens that way, particularly in relation to the relics of land animals which have to survive transportation from the hinterland down rivers to the sea for burial. They don't come complete with labels to say what they are. No bells ring to indicate which one is a dinosaur, so each fossil specimen has to be carefully extracted, studied and compared with other fossil material to try and find out what animal it came from.

For me, each specimen is a new challenge. Sometimes I'm right, often I'm wrong. The work is often tedious, the study difficult, the communication with overseas experts in museums and universities slow and expensive, but there is always enough excitement and stimulation to carry us along.

Since Joan Wiffen's first find, fossils from six different dinosaur species have been found in the Mangahouanga Stream. Wiffen, who continued her fossil-hunting habits well into her eighties, has received numerous awards and accolades for her work, including an honorary doctorate from Massey University and a CBE, both in 1994.

From *Valley of the Dragons: The Story of New Zealand's Dinosaur Woman* by Joan Wiffen: Random Century, Auckland, 1991.

Plastics that Conduct Electricity

In 2000, Alan Macdiarmid (1927–2007) became the third New Zealand scientist to win a Nobel Prize, when he shared the Prize in Chemistry with physicist Alan Heeger and polymer chemist Hideki Shirakawa for the discovery and development of 'conductive polymers', also known as synthetic metals. These new materials were remarkable because they combined the flexibility and malleability of a plastic with the electrical conductivity of a metal, and at a lower cost. This extract is from a short autobiography written for publication on Macdiarmid's receipt of the Nobel Prize. Alan Macdiarmid died in February 2007, as he was preparing to leave his Pennsylvania home to present his latest research to a nanotechnology conference in New Zealand.

My interest in chemistry was kindled when I was about ten years old, at which time I found one of my father's old chemistry textbooks dating back to the late 1800s when he was studying engineering. I spent hours poring over the pages in complete confusion but with burning curiosity! Some clarification of a type occurred when I rode my bicycle to the public library in Lower Hutt and entered the children's section. There, on the right-hand side on the bottom shelf, in the new books section, was a book with a bright blue cover. It was called *The Boy Chemist*. I took it out and continually renewed it by borrowing it for over a year and carried out most of the experiments in it.

When my father retired (on a very small pension) and moved away from Wellington, it was necessary for me to leave Hutt Valley High School after only three years at the age of sixteen and take a low-paying, part-time job as 'lab boy'/janitor in the chemistry department at Victoria University College, as it was then known. The total student population was 1,200; the chemistry department had a faculty of two! I boarded with friends of my parents and, as a part-time student, took only two courses—one in chemistry and one in mathematics. … One of my duties as lab boy, when I was not washing dirty labware or sweeping floors, was to prepare demonstration chemicals for Mr A. D. 'Bobbie' Monro, the lecturer in first-year chemistry. On one occasion he asked me to prepare some S_4N_4—beautiful bright orange crystals. When it became time for me to start my MSc thesis, I asked Mr Monro if I could look at some of its chemistry. He agreed. This resulted in my first publication in 1949. Its derivatives were highly coloured. Colour continued to be one of the driving forces in my future career in chemistry. I love colour. Little did I know that thirty years later this was going to be a key factor which would shape my professional life.

In 1950, I had the good fortune to receive a Fulbright fellowship from the US State Department to do a PhD at the University of Wisconsin, where I studied under Professor Norris F. Hall, majoring in inorganic chemistry, studying the rate of exchange in ^{14}C-tagged complex metal cyanides. It was at the University of Wisconsin that I became president of the International Club—the largest student organisation on campus—and had the crucial chance meeting of my life when I met my future wife, Marian Mathieu, at an International Club dance. During this time I was elected by the department of chemistry to the position of Knapp Research Fellow and had the privilege of living rent-free in the beautiful old ex-governor's mansion on the shores of Lake Mendota.

When I was still at the University of Wisconsin I was successful in obtaining a New Zealand Shell graduate scholarship to study silicon hydrides at Cambridge University, England under the directorship of Professor H.J. Emeléus. It was there that Marian and I were married in the chapel at my college, Sidney Sussex College.

After a brief appointment as a junior faculty member at Queen's College of the University of Saint Andrews, Scotland, I accepted a junior position on the faculty of the department of chemistry at the University of Pennsylvania, where I have been for the past 45 years, and became father of three daughters and a son and grandparent of nine lovely boys and girls. I grew to love teaching and the stimulation of young fresh inquiring minds. ...

I had the good fortune to meet my future friend and colleague Professor Alan Heeger, professor of physics at the University of Pennsylvania. On one occasion he came to my office and informed me that Mort Labes, professor of chemistry at Temple University, had published a paper on a highly conducting material. I asked Heeger its formula and he replied, 'ess-en-ex.' Being an inorganic chemist, I wrote down on a piece of paper, '$(Sn)_x$' and said, 'Of course you expect it to be conducting, it's a metal!' To which Heeger replied on paper, 'No, not $(Sn)_x$, but $(SN)_x$!' This was the beginning of our each learning each other's scientific language. I told him that I had made the precursor to $(SN)_x$—that is, S_4N_4— during my MSc thesis work in New Zealand. He asked me if I could make some $(SN)_x$—as golden crystals. We were ultimately successful, and co-published many papers together on this conducting polymer.

When I was a visiting professor at Kyoto University in Japan, lecturing on molecular silicon compounds, I visited Tokyo Institute of Technology in 1975 and described our work on $(SN)_x$. Hideki Shirakawa and I met over a cup of green tea after a lecture I gave, and as I was showing a sample of our golden $(SN)_x$ he showed me a sample of his silvery $(CH)_x$.

I asked him how he had made this silvery film of polyacetylene and he replied that it had occurred because of a misunderstanding between the Japanese language and that of a foreign student who had just joined his group. Shirakawa had been polymerizing ordinary acetylene welding gas using a Ziegler-Natta catalyst and had been obtaining a rather uninteresting blackbrown powder. He told the new student to repeat this work using a concentration of the catalyst which was milli-molar. A few days later the student came back and said that the stirring bar would not go around in the flask. Shirakawa went to the laboratory and, sure enough, instead of the black-brown powder, there were lumps of silvery-pinkish jelly floating around. Shirakawa asked what the student had done and the student replied that he had done exactly as Shirakawa had told him; he had made the catalyst with a concentration of 'x-molar'—in other words, he had made the catalyst 1,000 times more concentrated than Shirakawa had told him!

Shirakawa was most intrigued by this observation, since, as all good chemists know, a catalyst should only increase the rate of a chemical reaction and should not alter the nature of the product. This then started Shirakawa investigating this silvery form of polyacetylene. I asked Shirakawa if he could join me for a year at the University of Pennsylvania since I was already interested in conducting materials such as the golden $(SN)_x$ films. He stated that he could, and when he arrived we tried to make the silvery polyacetylene, $(CH)_x$, more pure, and hence increase its conductivity. However, we found that the purer we made the $(CH)_x$, by elemental analysis, the lower was its conductivity.

Since we had found previously that by adding bromine to the golden $(SN)_x$ material we could increase its conductivity tenfold, we thought that perhaps the impurity in the polyacetylene was acting as a dopant and was actually increasing the conductivity of the polyacetylene, rather than decreasing it. We therefore decided to add some bromine to the silvery $(CH)_x$ films and immediately, within a few minutes at room temperature, the conductivity increased many millions of times. We then collaborated with my colleague Professor Alan Heeger, who was well-versed in the physics of conducting materials. The rest is history!

From 'Autobiography' by Alan MacDiarmid, *Les Prix Nobel, The Nobel Prizes 2000*, edited by Tore Frängsmyr, Stockholm.

Bright Star

'If ever you thought scientists were unemotional impersonal eggheads, change your mind!' astronomer Beatrice Hill Tinsley (1941–81) wrote in 1974, in a letter to her sister Rowena. Hill Tinsley was living proof of her own statement, not only one of the most important and remarkable scientists New Zealand has ever produced but a woman of complex emotions, with a passion for astronomy, music and her family. After being educated in New Zealand, she studied and worked in the United States in the 1960s and '70s, where she proved that, rather than being static, the universe is expanding. And by showing that galaxies are not fixed but are changing and interacting with each other, she pioneered an entirely new field of astronomical research.

Hill Tinsley was a prolific and enthusiastic correspondent, and wrote regularly to her family in New Plymouth. In this letter from April 1958, before she was married, she tells of her life as a first-year student at the University of Canterbury.

Every day I get more and more thrilled with science. The other night at the Metcalfs* I spent ages talking with the Doc about chemistry and physics. Gosh he's an interesting person. He was telling me about various living scientists of most outstanding insight and ingenuity, who are working on things now and the work they've done—all revolutionary ideas and results. Dr M is totally fascinated by it all himself—it sort of pours from him as he speaks—and he talks and talks about all the discoveries and says, 'Isn't it thrilling?' I do agree! He sees science as it really is and not as a mere 'subject'.

---

* Dr Wal Metcalf was Hill Tinsley's chemistry lecturer.

I told him I thought it was a pity the way stage one had to skim over all sorts of atomic things or learn half-accurate explanations for them, and he quite agreed and told me all sorts of things about the way he had to adjust his lecturing technique so that he can teach oversimplified things without letting us think that they are sufficient.

I also let forth a lot of things about how maddening it was (a) to have to learn the dry sides of physics that would have no bearing on an atomic scientist's career, and (b) to have to do such ghastly easy maths—all of which he explained why it had to be so. He doesn't care in the least about degrees and exams etc.; he only cares about knowledge.

I'm therefore feeling very revolutionary and I'm damned sick of working for exams. Consequently I'm going to put first things first and instead of listening to Mr __ waffling through applied maths that I did last year, or __ mumbling thru' calculus as I did in 6b, I'm going to sit in the back in lectures and work on stuff that's going to get me somewhere, namely learning new maths. Also, instead of just sticking to the syllabus work in chemistry and physics, I'm reading and learning as much as I can about everything. Obviously there's no chance of doing any original work until one has a wide background of present knowledge and a very wide knowledge of maths.

So here I go. Even if I never achieve any of my aims, I'll at least have had the thrill of knowing where I want to go—along those lines, that is. And the same applies to all parts of life—it's the wanting that matters, and having a definite goal ahead. Even if one has about ten thrilling goals it hardly matters!

By 1965, Hill Tinsley was enrolled in a PhD at the University of Texas in Austin and was planning her thesis, which would eventually revolutionise astronomers' understanding of galactic evolution. In April of that year she wrote to her family to explain her research.

Really I will be studying a whole lot of different theories of cosmology, to see which is best able to explain the observations made with optical and radio telescopes on different galaxies. The theories are based on Einstein's General Relativity (that means Steady State is excluded) and they represent different motions of the galaxies in the expansion of the universe, and different ages of the universe, different number of galaxies per unit volume of space, and so on.

Using each theory one works out what to expect in observations of the faintest galaxies, quasars and radio sources. Then one tries to choose the theory that fits best. But I do not think the observations are complete enough yet for a real choice to be made.

The idea is to have a systematic and extensive set of calculations to compare with the observations as they become available, as it is a very interesting question which theory is best. An answer would have a lot of information about the past and future of the universe.

The calculations also depend on what the various galaxies etc. were like at the time the light left them, the light which now gets to the telescopes, and it isn't necessarily true that they were the same as nearby objects! So I will have to include a variety of different assumptions about the galaxies themselves, which will be a matter of astrophysics.

Dr Sachs suggests that a good systematic set of calculations is really needed, though lots of people have done bits of this before of course. He is very good at seeing things straight, and seeing the wood in the trees of calculations that have been done in the past; so I think I will have very good guidance from him.

Already it looks a very interesting subject, because there are serious difficulties in trying to make any of the theories fit! (In case Father remembers how fond people were of the Steady State idea several years ago, I should mention that there have been observations of radio sources which seem to have ruled it out entirely. Even Hoyle agrees.)

Of course if it is impossible to explain the observations with any of the General Relativity theories, that will be worth finding out … I am still up to my ears in all sorts of books on astronomy for the prelim, and wishing I could get on with my thesis because the plan of attack is clear now, and very intriguing.

> Beatrice Hill Tinsley completed her PhD in 1967 and wrote many detailed papers on this and other topics over the next decade. In 1974 she was appointed associate professor of astronomy at Yale University, where she realised her life's goal and became a cosmologist whose work was recognised and applauded throughout the scientific world. Sadly, this brilliant scientist died in the prime of her life and career. By 1980, when she wrote this letter to her father, she knew the cancer that had first emerged as a dark mole on her leg was going to kill her. She died in 1981, aged 40.

I've been one of the luckiest people … to realise my lifelong ambitions, and far more so than I could have hoped.

I have vivid memories from my early high school years of studying astronomy in the *Oxford Junior Encyclopaedia*, and wanting to contribute knowledge in cosmology in particular. Must admit I always wanted to be famous, as you and Mummy tell me from a young age! I have to thank you and Mummy for the total encouragement you always gave me; I always thought you believed that I could do practically anything I tried, and the difference that made in my life must be incalculable. And unlike so many 'ambitious' parents, you made it clear that the choice of what to do was my own. (I have since learnt how many parents discouraged daughters from science.) I owe you a vast amount in life!

From *Bright Star: Beatrice Hill Tinsley, Astronomer* by Christine Cole Catley: Cape Catley, Auckland, 2006.

Lovemaps!

John Money (1921–2006), a pioneering sexologist, began his career as a psychology lecturer at the University of Otago, where he taught and befriended a young Janet Frame. He went on to a career in the United States, and directorship of the Psychohormonal Research Unit at Johns Hopkins University in Baltimore.

As the historian Michael King wrote in a 1998 *Listener* article, John Money's career progress was meteoric. 'In 1951 he moved from Harvard to Johns Hopkins, where he founded the Psychohormonal Research Unit and, just over a decade later, the Gender Identity Clinic. He was the pioneer in the treatment of hermaphroditic or "intersex" children by programmes of sexual assignment, which often involved surgery, hormonal treatment and counselling. He was involved in sexual reassignment programmes for transsexuals. He coined and defined the now widely used terms "gender role", and "lovemap". He made major contributions to theoretical understandings of homosexuality, bisexuality and heterosexuality. And he devised successful psychological and hormonal programmes for the treatment of so-called sexual deviants, for whose often kinky disorders he revived the word "paraphilia". He taught generations of medical students at Johns Hopkins. And he has written or co-authored about forty books and nearly 400 scientific papers.'

While Money's belief that gender identity was largely learned, and that children younger than eighteen months could be 'assigned' to either gender, would prove controversial, other aspects of his work, such as the many words and phrases he has added to the psychologist's lexicon, are widely accepted and in common use. Here he writes, rather playfully, about 'lovemaps'.

Lovemaps! They're as common as faces, bodies and brains. Each of us has one. Without it there would be no falling in love, no mating, and no breeding of the species. Lacking a name, however, the lovemap has existed in a conceptually unexplored territory of the mind, unknown to science and scholarly inquiry.

By searching through my file of manuscripts, I found that I first wrote the word 'lovemap' in 1980. It was in an article titled 'Pairbonding and Limerence', published in 1983 ... Before I wrote that article, I had already begun to talk about lovemaps in my lectures to students whose textbook was *Love and Lovesickness: The Science of Sex, Gender Difference and Pairbonding* which the Johns Hopkins University Press had published for me in 1980. In that book I had written ... 'There is a rather sophisticated riddle about what a boyfriend (or girlfriend) and a Rorschach ink-blot have in common. The answer is that you project an image of your own on to each. In many instances, a person does not fall in love with a partner, per se, but with a partner as a Rorschach love-blot. That is to say, the person projects on to the partner an idealised and highly idiosyncratic image that diverges from the image of that partner as perceived by other people. Hence the popular idiom that love is blind, for a lover projects on to a partner, or love-blot, his/her unique love image, as unique as his/her own face or fingerprint.'

To communicate fluently with students, I found it extremely awkward to have only the expression 'an idealised and highly idiosyncratic image'. Therefore, I began substituting the single term 'lovemap'. Everyone knew immediately exactly what I was talking about. The word became part of my vocabulary not only with students but also with patients. Patients also knew, without hesitation, the meaning of the word. They were adept also at deciphering their own personal lovemaps and the errors, if any, in them.

People who have heard the title of this book, or who have read it in typescript, now include the word 'lovemap' in their everyday vocabulary. Sooner or later, therefore, 'lovemap' will find its way into the standard dictionaries of the English language, and in translations. For those who will need a definition: a lovemap is not present at birth. Like a native language, it differentiates within a few years thereafter. It is a developmental representation or template in your mind/brain, and is dependent on input through the special senses. It depicts your idealised lover and what, as a pair, you do together in the idealised, romantic, erotic and sexualised relationship. A lovemap exists in mental imagery first, in dreams and fantasies, and then may be translated into action with a partner or partners.

From *Lovemaps: Clinical Concepts of Sexual/Erotic Health and Pathology, Paraphilia, and Gender Transposition in Childhood, Adolescence, and Maturity* by John Money: Irvington Publishers, Inc., New York, 1986.

A Survivor from
the Dinosaur Age

The tuatara, a lizard-like reptile, belongs to one of New Zealand's most ancient lineages, the order Sphenodontia, which emerged 220 million years ago in the age of the dinosaurs. In 1990, zoologists Charles Daugherty (born 1946) and Alison Cree (born 1960) cemented the work of a five-year research programme by hatching thirty tuatara eggs of the rare Brothers Island population in incubators at the School of Biological Sciences at Victoria University of Wellington. Before their project, which they describe here, little was known about tuatara reproduction, but by 1995 the continued work of Cree (by now at the University of Otago), Daugherty and their Australian colleague Mike Thompson had led to the discovery that tuatara are sub-ject to a phenomenon known as temperature-dependent sex determination: tuatara hatched from eggs incubated at cool temperatures are always female, and tuatara from eggs incubated at warmer temperatures are male. Their numbers boosted by the captive breeding programme, tuatara now inhabit thirty-seven rat-free islands and the Karori Wildlife Sanctuary in Wellington.

Hatching is too gentle a word to describe the birth of a tuatara. Over a period of months, the soft-shelled tuatara egg absorbs moisture from the soil, swelling up like a balloon until it is a tight-skinned capsule. Then, using its egg tooth—a sharp-pointed spike on the end of its snout—the baby tuatara punctures the shell, and its wet head literally explodes into view. Over the next few hours a series of abrupt wriggling movements will free the hatchling from the egg that has been its home for the last twelve months.

Few people have observed the hatching of a tuatara, which usually occurs in a cool, dark nest about 15 centimetres below ground. But in May of this year, thirty tuatara eggs from North Brothers Island in Cook Strait were hatched in incubators at Wellington's Victoria University, giving us the opportunity to observe closely an event that has been happening for more than 200 million years.

Tuatara are the last surviving members of a lineage that stretches back to the Mesozoic—the beginning of the 'Age of Reptiles'. Their ancestors witnessed not only the immense, terrifying diversity of the dinosaurs, but also geological upheavals that shuffled the continents around the globe like jigsaw pieces. Perhaps they even watched from their burrows as the Earth shuddered under the impact of a giant meteorite—a disaster some scientists think occurred about 65 million years ago and led to the extinction of the dinosaurs.

Somehow the 'proto-tuatara' survived this cataclysm, hung on during the proliferation of birds and mammals, and eventually gave rise to the modern version, which survives only in New Zealand, and only just. Since humans arrived, about a thousand years ago, tuatara numbers have declined rapidly. They disappeared from the mainland a hundred years ago, and are now found only on a diminishing number of offshore refuges.

The birth in captivity of thirty healthy tuatara may mark a turning point in the long history of the animal, for these juveniles are destined to recolonise some of the islands from which tuatara have vanished. They are the culmination of a long-term research programme that we hope will turn the tide of fate in the tuatara's favour, away from an extinction that many have felt to be inevitable.

Tuatara have puzzled and fascinated scientists for more than a century. Since the late 1800s, naturalists have beaten a path to these shores to collect tuatara—sometimes hundreds at a time—for the world's museums. But studying pickled specimens doesn't save a species, and the emphasis in tuatara research is now on their behaviour and ecology.

During the last five years, teams of biologists and conservationists, working together in a project organised by Victoria University, have tried to answer basic questions about the tuatara's reproductive biology, social behaviour and genetics. Such research is not easy. Access to the islands on which tuatara occur has long been restricted by both weather and New Zealand law. That tuatara are mainly nocturnal and have a life span that is longer than the normal scientific career does not make the job any simpler.

Besides rock-climbing skills, a good pair of sea legs and the capacity to survive on dried foods for extended periods, all tuatara researchers require one crucial skill: the ability to catch the subjects of their study. Tuatara emerge at dusk from their burrows and spend most of the night near the burrow entrance, waiting for a tasty meal such as a large weta or lizard to wander within striking distance. Sometimes they forage away from the burrow, perhaps on sun-warmed rocks near the high tide mark, where lizards are also searching for a meal. If the tuatara is lucky, an unwary skink may soon become supper.

Spotlit by the beam of a torch, a tuatara will do one of two things: turn tail and scuttle down the nearest burrow, or freeze like a possum. Fortunately, most choose the second option. Then, cautiously, a quick grab around the neck, just behind the powerful jaws, and the capture is successful.

Some of us have learned the hard way that a moment's carelessness may exact a painful price: a bite from teeth perfected over tens of millions of years for grasping prey securely, crushing it in powerful jaws and shearing it apart even as it struggles to escape.

When a tuatara clamps its sharp teeth into your bare finger, the searing pain endures until the tuatara finally decides to let go—which may be many minutes, because the tuatara has nothing if not patience.

Even if we avoid being bitten, tuatara can be very difficult to hold. Large males, the biggest as long as your arm and weighing over a kilogram, can put up a real fight, clawing and thrashing and grunting fiercely. Equally often, however, the tuatara is almost docile, displaying a stoicism that seems somehow appropriate to its antiquity.

While tuatara may be comparatively fearless at night, they are secretive and extremely wary during the daytime. Perhaps it is because the danger from predators, especially harriers circling relentlessly above many islands, is greater during the day. As a consequence, they are difficult to catch, and the risks are considerably higher. Tuatara seldom venture far from the burrow entrance during daylight hours. Usually, only the head is seen—especially if a ray of sunlight can warm it. But it's nearly impossible to sneak up and grab the animal before it swiftly retreats to safety down a burrow that may extend five or more metres below ground.

Often, no tuatara at all can be seen during the day. Then, the only way to find, and possibly catch, one is to go in after it. Lying face down on the fine, bare, cold soil, you slowly reach your arm into the burrow, searching blindly and gingerly for the soft skin of a tuatara—and often hoping you find nothing at all!

The word 'tuatara' means 'spinyback' in Maori. The spines, like the skin, are surprisingly soft, much like cool, dusty linen cloth to the touch, and pose no threat. If you are lucky, your hand lands on an exposed tail or leg or, best of all, the spines and back. Then, you press the tuatara firmly to the ground, securing it until you get a grip strong enough to pull the reluctant reptile out of the burrow. If unlucky, you find nothing, or you may just touch a tuatara as it retreats beyond your reach. Or, something may grab you. Petrels and shearwaters often use the same burrows as tuatara, and some of them have strong bites if disturbed. Worse, large centipedes with painful, poisonous stings also share burrows with birds and tuatara on islands in the Hauraki Gulf and the Bay of Plenty. Or, a large tuatara may express its displeasure with a sharp, unrelenting bite.

Bitten or not, a visit to a tuatara island is the experience of a lifetime. Few people ever get the opportunity, though, because tuatara have been fully protected since 1895, and permits to land on their home turf are not given lightly. But what astonishing places these islands are! Many, like Stephens Island in Cook Strait or Tawhiti Rahi and Aorangi Islands in the Poor Knights, rise straight up from the sea like stark, primeval fortresses. Cliffs a hundred metres high are topped with thick carpets of wind-shaped scrub: taupata, ngaio, or mahoe. In spring, the northern islands are red with flowering pohutukawa and, on a few islands, the glorious Poor Knights Lily.

The surrounding seas are often turbulent, and landing a dinghy can be both difficult and dangerous; on many of these islands there is no such thing as a beach. A research team of four to six people must be shuttled ashore, the dinghy stacked high above the gunwales with supplies. There's no going back for forgotten gear, so everything required for a week's work has to be remembered and included: food, tents, personal gear, scientific supplies, and plenty of torches and batteries, because most work is done at night. Often, all our drinking water has to be taken with us, too.

Once ashore, several hours are spent drying out wave-swamped gear, pitching tents in tiny openings at the margin of the scrub, and preparing gear for the first night's work. In daylight, tuatara islands often seem strangely quiet, with only a few parakeets or bellbirds for company, so the first few hours ashore are a peaceful interlude before the real work—and excitement—begins.

Signs of life are everywhere. The soil is usually bare from continual digging and trampling by birds and tuatara, and is riddled with burrows. But the scrub is thick, and we often have to crawl under or through the brittle, scratching branches. The combination of forest, soil and bird droppings gives the area a distinctive, pungent bouquet. Particularly ripe is the smell of a blue penguin nest, full of decaying faeces and the remnants of fish and squid regurgitated by the parents as food for the young.

As night approaches, there is an explosion of life. Most tuatara islands are free from mammals, and thus teem with birds, lizards, weta and beetles. From August to November, when seabirds return for breeding, the night-time cacophony of tens of thousands of these creatures creates a wall of noise. Sleep is almost impossible, not only because of the din, but because petrels crash-land on your tent with monotonous—but nevertheless startling—regularity. Then, just as you manage to doze off with the approaching dawn, an army of raucous penguins marches past your tent on their way to the sea!

Walking around these islands at night can be unnerving, as some seabirds seem attracted to the headlamps we wear for illumination. Every few steps a surprised bird bounces off an equally surprised scientist. On islands in the Hauraki Gulf, the beaches are alive with lizards, too—black Suter's skink and the brown, velvety Duvaucel's gecko, the largest surviving gecko in New Zealand. Under the low forest canopy on several islands, giant weta prowl through the vegetation, careful to stay out of the way of tuatara.

Tuatara islands differ dramatically from the New Zealand that humans have created with their cities, highways, farms, orchards, exotic forests, rabbits and hedgehogs. The forests are silent because the birds that once teemed in them have been killed by rats, cats, and stoats; the trees themselves dying because of possums brought from Australia. To visit a tuatara island is to travel backwards in time for a thousand years, or a million, or ten million—to a time when most of New Zealand shared the extraordinary biological diversity now found only on those few offshore islands where introduced mammals are absent.

Some species are gone forever, of course: moa, sea eagles, the giant gecko. But on tuatara islands, life is super-abundant. Seabirds fertilise the soil with their droppings, producing the rich plant communities that in turn provide food for insects, lizards, forest birds, and, at the top of the food chain, tuatara.

On Stephens Island in Cook Strait, the average tuatara weighs 400 to 500 grams, and in some places as many as two thousand tuatara share one hectare of forest—almost a tonne of tuatara per hectare. Even in poorer habitats, numbers are as high as five hundred per hectare. Such numbers are possible only because the soil, enriched by the tens of thousands of fairy prions that return to Stephens each year to breed, supports a diverse biological community that tuatara see as an enormous buffet.

Anything that moves is fair game to a tuatara: earthworms, beetles, lizards (seven species on Stephens Island), frogs, weta, injured or juvenile prions, and even, as our colleague Mary McIntyre discovered, young tuatara. Mary was studying hatchling and juvenile tuatara on Stephens Island. The behaviour and ecology of young tuatara have long been a mystery, because they are seen so infrequently, even on islands where adult numbers are high. To discover where the young hide, Mary taped small spools of cotton thread to the tails of a few juveniles and tied the ends of the threads to nearby plants. As the juvenile moved about, the thread unwound behind it, leaving a complete track of all its movements. Twice daily, she would begin at the tied end and follow each animal, mapping its entire path, finally clipping off the unwound thread when she caught up to the young one, and starting the process over again.

The tuatara is the sole living representative of the order Sphenodontia. With a lineage stretching back some 225 million years, tuatara—often referred to a 'living fossils'—are now found only in New Zealand. Dave Hansford/Hedgehog House

Day after day, Mary followed juvenile tuatara this way, showing that they sought shelter from the hot summer sun under small rocks in open areas, under thick clumps of grass, even under the leaves of thistles, and along the forest margin. Unexpectedly, however, they were most likely to move about in the daytime, despite the danger of overheating or drying out.

The reason soon became apparent: 'I followed the thread trail of a year-old tuatara that disappeared under a rock in the lighthouse keeper's sheep pasture. Gently lifting the rock to find the juvenile, I was shocked to see the thread disappearing into the mouth of a large adult male along with the tip of the tail of the young one.'

Mary concluded that adult tuatara are probably an important predator of baby tuatara, which may explain why juveniles are most active in the daytime, seeking shelter at night when adults are foraging for food.

Studies of tuatara on Stephens Island began in earnest over forty years ago. Bill Dawbin, then a lecturer at Victoria University specialising in whale biology, visited Stephens with an American herpetologist. Dawbin was captivated by what he saw, and returned repeatedly between 1949 and 1981 with teams of field workers who marked and measured hundreds of tuatara each trip. The capture and measurement of the same individuals over a thirty-year period allowed Dawbin and others to confirm that tuatara are extremely slow-growing, reaching reproductive maturity between ages ten to fifteen, continuing to grow until age thirty, and living for at least sixty to seventy years. Other studies during the 1970s and early 1980s by Don Newman and Geoff Walls, both now with the Department of Conservation, provided important additional information about numbers, food, and territory size of tuatara on Stephens.

A new set of studies began in 1985, when Mike Thompson, a tall, gregarious Australian, arrived at Victoria University. An expert in the nesting ecology and egg incubation of turtles, Mike quickly convinced us that similar knowledge for tuatara was essential for their preservation.

The first problem was finding tuatara nests. Previous reports of nesting were scarce, at best anecdotes that required confirmation. Even the timing of nesting was uncertain. Don Newman had taken X-rays of tuatara, indicating that eggshell formation occurred as early as September, but some females were obviously still carrying eggs as late as December. Mike began intensive searches for nesting females in September 1985, continuing until the end of November. None was found.

The next year, Mike assembled a small army of volunteers (including one of us—Alison Cree) to conduct nightly patrols for nesting females on Stephens Island from September until December. Team members searched either for females digging burrows or for 'scrapes', evidence of such digging. For weeks, we followed numerous female tuatara daily, recording their every movement, but finding no evidence of nesting.

At last, in November 1986, the action we had long anticipated was discovered. Several females that had not moved more than a few metres from their home burrows since September suddenly sprinted several hundred metres in a few nights, disappearing down cliff faces too steep for us to follow. Later that month, other nest sites were found, but in a surprising location. In the previous year, searches for nesting females had focused on the small patches of remnant forest on top of the island. In 1986, Mike established a search pattern that covered not only the forest areas, but also the expanses of pasture used for the past century by lighthouse keepers to raise sheep. (Most of Stephens Island was cleared of forest last century, making it the least natural of any tuatara island.)

On a warm night in November 1986, Mike stumbled on a nesting tuatara that was within sight of the front window of the field station. A rapid search revealed that females were nesting in rookeries all over the sheep pasture. Months of despondency changed to instant elation, and we immediately began marking the locations of all nest sites, using metre-high stakes topped with reflective tape. …

It turns out that the nesting behaviour of tuatara can be easily observed, once you know where to look. Apart from sea turtles that come ashore in large numbers to nest on sandy beaches, most reptiles are very secretive during egg-laying. Female tuatara, however, have prolonged, highly visible and surprisingly active nesting behaviour. Mike Thompson describes it: 'They come to nesting areas from hundreds of metres away, spend several days or even weeks digging a shallow nest hole, and then lay about ten eggs in it. Once the eggs are laid, the female fills the nest hole with soil and grass, returning to it nightly for up to a week after laying. Females even appear to guard the nest, presumably from other females, who have been seen digging up other females' nest holes to use for their own eggs.'

For the next three years, Mike and Alison returned monthly to check on the development of nests and to study the reproductive cycle of adult tuatara. Some of the animals studied were, in fact, the same animals marked decades earlier by Bill Dawbin and his teams. Many had not grown in the intervening period and were probably at least sixty to seventy years old. Scientists of the future will check on the survival of these well-known animals, and when such studies are made, we may discover that a few of the biggest, oldest tuatara on Stephens Island watched Captain Cook sail by in 1769.

From 'Tuatara: A Survivor from the Dinosaur Age' by Charles Daugherty and Alison Cree in *New Zealand Geographic*, April–June 1990.

Tuatara

Tuatara populations are being re-established on offshore islands, including Matiu/Somes Island in Wellington Harbour, Te Whanganui a Tara. In 1998, fifty-four tuatara were transferred to Matiu/Somes Island from Brothers Island and they are now breeding. Lucky visitors may occasionally come across a tuatara basking in the sun. This poem is by Wellington poet Nola Borrell.

Keep your distance
you're new here
rough-edged and arrogant

One step closer
and you won't see me
you won't see me anywhere

Always lie low, I say
I've learnt a thing or two
over 200 million years

Take away your 'ecologically
appropriate quarters'
this drainpipe will do

And quit drooling over me
I pounce on skinks and wetas
eat my own kind

If a female won't dance
I bite her neck
and then I take her

Still, I'm glad of a bit of
company. Keep our numbers up
and keep yours down

Not that I'm worried
I've outlived the dinosaur
I may outlive you

How many years did you say you'd been here?

From *Turbine 01*: www.victoria.ac.nz/turbine/borrell.html.

The Recent African Genesis of Humans

Evolutionary biologist Allan Wilson (1934–91) was born in Pukekohe and completed a degree at the University of Otago before leaving for post-graduate study in the United States, where he remained. While working at the University of California at Berkeley in 1967, Wilson attracted great controversy by arguing that the origins of the human species could be dated through use of a 'molecular clock'. Human evolution had previously been charted solely through use of fossils, but Wilson argued that if two species, for example humans and chimpanzees, had a common ancestor, the DNA of each species would have accumulated a steady rate of random mutations since their split. This technique was then used to date the human–chimpanzee split as having happened only five to seven million years ago. Twenty years later, Wilson used his molecular clock theory to calculate that humankind had a common maternal ancestor, 'Eve', who lived in Africa only 150,000 to 200,000 years ago.

While Wilson's ideas were radical, and often ridiculed at the time, his theories of the molecular clock and 'mitochondrial Eve' are now widely accepted. This 1992 article from *Scientific American* was co-written by Wilson and Rebecca Cann (born 1951), a former PhD student of his at the University of California, Berkeley.

In the quest for the facts about human evolution, we molecular geneticists have engaged in two major debates with the palaeontologists. Arguing from their fossils, most palaeontologists had claimed the evolutionary split between humans and the great apes occurred as long as 25 million years ago. We maintained human and ape genes were too similar for the schism to be more than a few million years old. After fifteen years of disagreement, we won that argument, when the palaeontologists admitted we had been right and they had been wrong.

Once again we are engaged in a debate, this time over the latest phase of human evolution. The palaeontologists say modern humans evolved from their archaic forebears around the world over the past million years. Conversely, our genetic comparisons convince us that all humans today can be traced along maternal lines of descent to a woman who lived about 200,000 years ago, probably in Africa. Modern humans arose in one place and spread elsewhere.

Neither the genetic information of living subjects nor the fossilised remains of dead ones can explain in isolation how, when and where populations originated. But the former evidence has a crucial advantage in determining the structure of family trees: living genes must have ancestors, whereas dead fossils may not have descendants. Molecular biologists know the genes they are examining must have been passed through lineages that survived to the present; palaeontologists cannot be sure that the fossils they examine do not lead down an evolutionary blind alley.

The molecular approach is free from several other limitations of palaeontology. It does not require well-dated fossils or tools from each part of the family tree it hopes to describe. It is not vitiated by doubts about whether tools found near fossil remains were in fact made and used by the population those remains represent. Finally, it concerns itself with a set of characteristics that is complete and objective.

A genome, or full set of genes, is complete because it holds all the inherited biological information of an individual. Moreover, all the variants on it that appear within a population—a group of individuals who breed only with one another—can be studied as well, so specific peculiarities need not distort the interpretation of the data. Genomes are objective sources of data because they present evidence that has not been defined, at the outset, by any particular evolutionary model. Gene sequences are empirically verifiable and not shaped by theoretical prejudices.

The fossil record, on the other hand, is infamously spotty because a handful of surviving bones may not represent the majority of organisms that left no trace of themselves. Fossils cannot, in principle, be interpreted objectively: the physical characteristics by which they are classified necessarily reflect the models the palaeontologists wish to test. If one classifies, say, a pelvis as human because it supported an upright posture, then one is presupposing that bipedalism distinguished early hominids from apes. Such reasoning tends to circularity. The palaeontologist's perspective therefore contains a built-in bias that limits its power of observation.

As such, biologists trained in modern evolutionary theory must reject the notion that the fossils provide the most direct evidence of how human evolution actually proceeded. Fossils help to fill in the knowledge of how biological processes worked in the past, but they should not blind us to new lines of evidence or new interpretations of poorly understood and provisionally dated archaeological materials.

All the advantages of our field stood revealed in 1967, when Vincent M. Sarich, working in Wilson's laboratory at the University of California at Berkeley, challenged a fossil primate called *Ramapithecus*. Palaeontologists had dated its fossils to about 25 million years ago. On the basis of the enamel thickness of the molars and other skeletal characteristics, they believed that *Ramapithecus* appeared after the divergence of the human and ape lineages and that it was directly ancestral to humans.

Sarich measured the evolutionary distance between humans and chimpanzees by studying their blood proteins, knowing the differences reflected mutations that have accumulated since the species diverged. ... To check that mutations had occurred equally fast in both lineages, he compared humans and chimpanzees against a reference species and found that all the genetic distances tallied.

Sarich now had a molecular clock; the next step was to calibrate it. He did so by calculating the mutation rate in other species whose divergences could be reliably dated from fossils. Finally, he applied the clock to the chimpanzee–human split, dating it to between five and seven million years ago—far later than anyone had imagined.

At first, most palaeontologists clung to the much earlier date. But new fossil finds undermined the human status of *Ramapithecus*: it is now clear *Ramapithecus* is actually *Sivapithecus*, a creature ancestral to orangutans and not to any of the African apes at all. Moreover, the age of some sivapithecine fossils was downgraded to only about six million years. By the early 1980s, almost all palaeontologists came to accept Sarich's more recent date for the separation of the human and ape lines. ...

Two novel concepts emerged from the early comparisons of proteins from different species. One was the concept of inconsequential, or neutral, mutations. Molecular evolution appears to be dominated by such mutations, and they accumulate at surprisingly steady rates in surviving lineages. In other words, evolution at the gene level results mainly from the relentless accumulation of mutations that seem to be neither harmful nor beneficial. The second concept, molecular clocks, stemmed from the observation that rates of genetic change from point mutations (changes in individual DNA base pairs) were so steady over long periods that one could use them to time divergences from a common stock.

We could begin to apply these methods to the reconstruction of later stages in human evolution only after 1980, when DNA restriction analysis made it possible to explore genetic differences with high resolution. Workers at Berkeley, including Wes Brown, Mark Stoneking and us, applied the technique to trace the maternal lineages of people sampled from around the world.

The DNA we studied resides in the mitochondria, cellular organelles that convert food into a form of energy the rest of the cell can use. Unlike the DNA of the nucleus, which forms bundles of long fibres, each consisting of a protein-coated double helix, the mitochondrial DNA comes in small, two-strand rings. Whereas nuclear DNA encodes an estimated 100,000 genes[*]—most of the information needed to make a human being—mitochondrial DNA encodes only thirty-seven. In this handful of genes, every one is essential: a single adverse mutation in any of them is known to cause some severe neurological diseases.

For the purpose of scientists studying when lineages diverged, mitochondrial DNA has two advantages over nuclear DNA. First, the sequences in mitochondrial DNA that interest us accumulate mutations rapidly and steadily, according to empirical observations. Because many mutations do not alter the mitochondrion's function, they are effectively neutral, and natural selection does not eliminate them.

This mitochondrial DNA therefore behaves like a fast-ticking clock, which is essential for identifying recent genetic changes. Any two humans chosen randomly from anywhere on the planet are so alike in most of their DNA sequences that we can measure evolution in our species only by concentrating on the genes that mutate fastest. Genes controlling skeletal characters do not fall within this group.

Second, unlike nuclear DNA, mitochondrial DNA is inherited

---

[*] The Human Genome Project, completed in 2003, would reveal there are only 20,000 to 25,000 genes in human DNA.

from the mother alone, unchanged except for chance mutations. The father's contribution ends up on the cutting-room floor, as it were. The nuclear genes, to which the father does contribute, descend in what we may call ordinary lineages, which are of course important to the transmission of physical characteristics. For our studies of modern human origins, however, we focus on the mitochondrial, maternal lineages.

Maternal lineages are closest among siblings because their mitochondrial DNA has had only one generation in which to accumulate mutations. The degree of relatedness declines step by step as one moves along the pedigree, from first cousins descended from the maternal grandmother, to second cousins descended from a common maternal great-grandmother and so on. The farther back the genealogy goes, the larger the circle of maternal relatives becomes, until at last it embraces everyone alive.

Logically, then, all human mitochondrial DNA must have had an ultimate common female ancestor. But it is easy to show she did not necessarily live in a small population or constitute the only woman of her generation. Imagine a static population that always contains fifteen mothers. Every new generation must contain fifteen daughters, but some mothers will fail to produce a daughter, whereas others will produce two or more. Because maternal lineages die out whenever there is no daughter to carry on, it is only a matter of time before all but one lineage disappears. In a stable population the time for this fixation of the maternal lineage to occur is the length of a generation multiplied by twice the population size.

One might refer to the lucky woman whose lineage survives as Eve. Bear in mind, however, that other women were living in Eve's generation and that Eve did not occupy a specially favoured place in the breeding pattern. She is purely the beneficiary of chance. Moreover, if we were to reconstruct the ordinary lineages for the population, they would trace back to many of the men and women who lived at the same time as Eve. Population geneticists … estimate that as many as 10,000 people could have lived then. The name 'Eve' can therefore be misleading—she is not the ultimate source of all the ordinary lineages, as the biblical Eve was.

From mitochondrial DNA data, it is possible to define the maternal lineages of living individuals all the way back to a common ancestor. In theory, a great number of different genealogical trees could give rise to any set of genetic data. To recognise the one that is most probably correct, one must apply the parsimony principle, which requires that subjects be connected in the simplest possible way. The most efficient hypothetical tree must be tested by comparison with other data to see whether it is consistent with them. If the tree holds up, it is analysed for evidence of the geographic history inherent in elements.

In 1988 Thomas D. Kocher of Berkeley … applied just such a parsimonious interpretation to the interrelatedness of the mitochondrial DNA of fourteen humans from around the world. He determined thirteen branching points were the fewest that could account for the differences he found. Taking the geographic considerations into account, he then concluded that Africa was the ultimate human homeland: the global distribution of mitochondrial DNA types he saw could then be explained most easily as the result of no more than three migration events to other continents.

A crucial assumption in this analysis is that all the mitochondrial lineages evolve at the same rate. For that reason, when Kocher conducted his comparison of the human mitochondrial DNAs, he also included analogous sequences from four chimpanzees. If the human lineages had differed in the rate at which they accumulated mutations, then some of the fourteen human sequences would be significantly closer or farther away from the chimpanzee sequences than others. In fact, all fourteen human sequences are nearly equidistant from the chimpanzee sequences, which implies the rates of change among humans are fairly uniform.

The chimpanzee data also illustrated how remarkably homogeneous humans are at the genetic level: chimpanzees commonly show as much as ten times the genetic variation as humans. That fact alone suggests that all of modern humanity sprang from a relatively small stock of common ancestors.

Working at Berkeley with Stoneking, we expanded on Kocher's work by examining a larger genealogical tree made up from 182 distinct types of mitochondrial DNA from 241 individuals. The multiple occurrences of mitochondrial DNA types were always found among people from the same continent and usually in persons who lived within 100 miles of one another. Because the tree we constructed had two main branches, both of which led back to Africa, it, too, supported the hypothesis that Africa was the place of origin for modern humans.

One point that jumps out of our study is that although geographic barriers do influence a population's mitochondrial DNA, people from a given continent do not generally all belong to the same maternal lineage. The New Guineans are typical in this respect. Their genetic diversity had been suspected from linguistic analyses of the remarkable variety of language families — generally classified as Papuan—spoken on this one island … On our genealogical tree, New Guineans showed up on several different branches, which proved that the common female ancestor of all New Guineans was not someone in New Guinea. The population of New Guinea must have been founded by many mothers whose maternal lineages were most closely related to those in Asia.

That finding is what one would expect if the African origin hypothesis were true: as people walked east out of Africa, they would have passed through Asia. Travel was probably slow, and during the time it took to reach New Guinea, mutations accumulated both in the lineages that stayed in Asia and in those that moved on.

Thus, people who are apparently related by membership in a common geographic race need not be very closely related in their mitochondrial DNA. Mitochondrially speaking, races are not like biological species. We propose that the anatomic characteristics uniting New Guineans were not inherited from the first settlers. They evolved after people colonised the island, chiefly as the result of mutations in nuclear genes spread by sex and recombination throughout New Guinea. Similarly, the light skin colour of many whites is probably a late development that occurred in Europe after that continent was colonised by Africans.

During the early 1980s, when we were constructing our genealogical tree, we had to rely on black Americans as substitutes for Africans, whose mitochondrial DNA was difficult to obtain in the required quantities. Fortunately, the recent development of a technique called the polymerase chain reaction has eliminated that constraint. The reaction makes it possible to duplicate DNA sequences easily, ad infinitum; a small starting sample of DNA can expand into an endless supply...

The polymerase chain reaction enabled Linda Vigilant ... to redo our study using mitochondrial DNA data from 120 Africans, representing six diverse parts of the sub-Saharan region. Vigilant traced a genealogical tree whose fourteen deepest branches lead exclusively to Africans, and whose fifteenth branch leads to both Africans and non-Africans. The non-Africans lie on shallow secondary branches stemming from the fifteenth branch. Considering the number of African and non-African mitochondrial DNAs surveyed, the probability that the fourteen deepest branches would be exclusively African is one in 10,000 for a tree with this branching order.

Satoshi Horai and Kenji Hayasaka of the National Institute of Genetics in Japan analogously surveyed population samples that included many more Asians and individuals from fewer parts of Africa; they, too, found that the mitochondrial lineages led back to Africa. We estimate the odds of their arriving at that conclusion accidentally were only four in 100. Although these statistical evaluations are not strong or rigorous tests, they do make it seem likely that the theory of an African origin for human mitochondrial DNA is now fairly secure.

Because our comparisons with the chimpanzee data showed the human mitochondrial DNA clock has ticked steadily for millions of years, we knew it should be possible to calculate when the common mother of humanity lived. We assumed the human and chimpanzee lineages diverged five million years ago, as Sarich's work had shown. We then calculated how much humans had diverged from one another relative to how much they had diverged from chimpanzees—that is, we found the ratio of mitochondrial DNA divergence among humans to that between humans and chimpanzees.

Using two different sets of data, we determined the ratio was less than 1:25. Human maternal lineages therefore grew apart in a period less than 1/25th as long as five million years, or less than 200,000 years. With a third set of data on changes in a section of the mitochondrial DNA called the control region, we arrived at a more ancient date for the common mother. That date is less certain, however, because questions remain about how to correct for multiple mutations that occur within the control region.

One might object that a molecular clock known to be accurate over five million years could still be unreliable for shorter periods. It is conceivable, for example, that intervals of genetic stagnation might be interrupted by short bursts of change when, say, a new mutagen enters the environment, or a virus infects the germ-line cells, or intense natural selection affects all segments of the DNA. To rule out the possibility that the clock might run by fits and starts, we ran a test to measure how much mitochondrial DNA has evolved in populations founded at a known time.

The aboriginal populations of New Guinea and Australia are estimated to have been founded less than 50,000 to 60,000 years ago. The amount of evolution that has since occurred in each of those places seems about one-third of that shown by the whole human species. Accordingly, we can infer that Eve lived three times 50,000 to 60,000 years ago, or roughly 150,000 to 180,000 years ago. All our estimates thus agree the split happened not far from 200,000 years ago.

Those estimates fit with at least one line of fossil evidence. The remains of anatomically modern people appear first in Africa, then in the Middle East and later in Europe and east Asia. Anthropologists have speculated that in east Africa the transition from anatomically archaic to modern people took place as recently as 130,000 years ago...

On the other hand, a second line of evidence appears to conflict with this view. The fossil record shows clearly that the southern parts of Eurasia were occupied by archaic people who had migrated from Africa to Asia nearly a million years ago. Such famous fossils as Java Man and Beijing Man are of this type. This finding and the hypothesis that the archaic Eurasian population underwent anatomic changes that made them resemble more modern people led to the multiregional evolution model: similar evolutionary changes in separate geographic regions converted the inhabitants from archaic small-brained to modern big-brained types.

Huge levels of gene flow between continents, however, would be necessary to maintain human populations as one biological species. The multiregional evolution model also predicts that at least some genes in the modern east-Asian population would be linked more closely to those of their archaic Asian predecessors than to those of modern Africans. We would expect to find deep lineages in Eurasia, especially in the Far East. Yet surveys in our laboratories and in others, involving more than 1,000 people from Eurasia and its mitochondrial DNA satellites (Australia, Oceania and the Americas), have given no hint of that result.

It therefore seems very unlikely that any truly ancient lineages survive undetected in Eurasia. We simply do not see the result predicted by the regional model. Moreover, geneticists such as Masatoshi Nei of Pennsylvannia State University, Kenneth K. Kidd of Yale University, James Wainscoat of the University of Oxford and Luigi L. Cavalli-Sforza of Stanford University have found support for an African origin model in their studies of nuclear genes.

Proponents of the multiregional evolution model emphasise they have documented a continuity of anatomic morphologies between the archaic and modern residents of different regions; they insist these morphologies would be unlikely to evolve independently in any invading people. For that argument to hold true, it must also be shown that the cranial features in question are truly independent of one another—that is, that natural selection would not tend to favour certain constellations of functionally related features. Yet we know powerful jaw muscles may impose changes on the mandible, the brow ridge and other points on the skull; circumstances that promoted the evolution of these features in one population might do so again in a related population.

Other palaeontologists also dispute the evidence for continuity. They argue modern populations are not linked to past ones by morphological characteristics that evolved uniquely in the fossil record. Instead, fossils and modern populations are united by their shared retention of still older ancestral characteristics. The continuity seen by believers in multiregional evolution may be an illusion.

The idea that modern humans could cohabit a region with archaic ones and replace them completely without any mixture may sound unlikely. Nevertheless, some fossil finds do support the idea. Discoveries in the caves at Qafzeh in Israel suggest Neanderthals and modern humans lived side by side for 40,000 years, yet they left little evidence of interbreeding.

How one human population might have replaced archaic humans without any detectable genetic mixing is still a mystery. One of us (Cann) suspects infectious diseases could have contributed to the process by helping to eliminate one group. Cavalli-Sforza has speculated the ancestors of modern humans may have developed some modern trait, such as advanced language skills, that effectively cut them off from breeding with other hominids. This and related questions may yield as molecular biologists learn how to link specific genetic sequences to the physical and behavioural traits those sequences influence.

Even before then, further studies of both nuclear and mitochondrial DNA will render more informative genetic trees. Particularly enticing are the sequences on the Y chromosome that determine maleness and that are therefore inherited from the father alone. Gerad Lucotte's laboratory at Collège de France has indirectly compared such sequences in an effort to trace paternal lineages to a single progenitor—'Adam', if you will. Those preliminary results also point to an African homeland, and with further refinements this work on paternal lineages may be able to provide an invaluable check on our results for maternal lineages. Unfortunately, base changes accumulate slowly on useful regions of the Y chromosome, making it technically difficult to conduct a detailed genealogical analysis.

Still more progress can be expected in the immediate future, as molecular biologists learn to apply their techniques to materials uncovered by our friendly rivals, the palaeontologists. Preliminary molecular studies have already been conducted on DNA from mummified tissues found in a Florida bog and dated to 7,500 years ago. Improved methods of extracting DNA from still older fossilised bone now appear close at hands. With them, we may begin building the family tree from a root that was alive when the human family was young.

From 'The Recent African Genesis of Humans' by Allan C. Wilson and Rebecca L. Cann in *Scientific American*, April 1992.

Ghosts of Gondwana

How did New Zealand come to be populated by such an unusual mixture of plants and animals? For a long time, the question was answered by the Moa's Ark theory, which supposed that the ancestors of New Zealand's current species were present on the Zealandia continent when it separated from the ancient supercontinent of Gondwana some eighty million years ago. However, as entomologist and natural historian George Gibbs (born 1937) describes here, recent molecular studies are providing evidence that dispersal across oceans may better explain the origin of many of our species.

Jared Diamond is probably best known for his book *Guns, Germs and Steel*, in which he sums up the landmark events in human history over the last 13,000 years. In 1990 he visited New Zealand for a conference on islands and conservation. In the course of a keynote address on the role of the New Zealand offshore islands for conserving endangered wildlife, he observed that 'New Zealand is as close as we will get to the opportunity to study life on another planet.' When the fledgling Zealandia land mass split away from the shores of Australis about 80 million years ago, a new opportunity for the evolution of life was about to begin. Unfettered by some of the events on the rest of the Earth's land masses, Zealandia was free to develop characteristics of its own. We really have very little knowledge of the flora and fauna of that new land, but we can study the outcome today and, from the windows we have on the journey along the way, we can fill in some of the important details of this story. It is a story of separation from the rest of the world—separation at a time when two groups of organisms that were destined to dominate the flora and fauna all over the world were just beginning their conquest: the flowering plants and the mammals.

The plant group was a little ahead of the animal group when Zealandia was born, and was probably more suited to trans-oceanic dispersal, with the result that they overwhelmed the earlier plants and dominated New Zealand in much the same way as they have elsewhere in the world. However, the animal group that did so well everywhere else failed to establish in New Caledonia or New Zealand, leaving these lands in the unique position of being mammal-free during the great mammal explosion. As Tim Flannery has pointed out, these islands were left with a natural experiment in which birds could call the tune.

Why did mammals fail to colonise Zealandia or, if they did reach it, why did they not succeed the way they did elsewhere? The answer to this riddle is not as simple as some might think. The prevailing view is that early mammals were simply not in the right place at the right time. They missed the boat. But there could be more to our mammal-free existence than that. We know that the most basal of all surviving mammals, the egg-laying monotremes, have a 120-million-year history in neighbouring Australia, and that their fossils are known from Patagonia and Antarctica. The sole surviving monotremes are the Australian echidna and platypus, and there is every reason to believe that their ancestors should have been part of the founding fauna of Zealandia. Whatever happened to them? And what about the marsupial mammals that became such a dominant force in Australia? There is now evidence that their closest ancestor is the diminutive South American monito del monte, *Dromiciops australis* … which lives in the southern beech forests of Chile. Since the marsupials of South America appear to be more ancient than those of Australia, the evidence suggests that the founding Australian marsupials dispersed from Patagonia across Antarctica prior to the separation of Australia from Antarctica. If they passed along this route some time after 80 million years ago, then these roving Antarctic marsupials would indeed have missed the opportunity to get to New Zealand.

If the monotreme mammals were around at the right time but failed to make it to New Zealand, and the marsupial mammals missed the departure date, where does that leave the placental mammals that dominate mammalian life everywhere except in Australia? Perhaps (apart from bats and seals) they missed out completely on access to the southern continent of Australis. But fossils from Seymour Island on the Antarctic Peninsula have revealed that both marsupial and placental mammals existed there about 40 million years ago. It has been argued that these placentals must have reached Antarctica from South America after the route to Australia was broken, and thus could play no part in the Australian fauna. However, an unresolved complication arose recently when 16-million-year-old vertebrate fossil material from St Bathans in Central Otago ... was found to contain small mammalian bones that were not bats'. Although their significance has yet to be determined, one thing is certain: the absence of grazing and predaceous land mammals had far-reaching effects on the evolutionary patterns in New Zealand. ...

The original land of Zealandia must have been densely populated with plants and animals when it broke away. ... The land area was greater and the climate warmer than that of New Zealand today, so it is likely the biota, amounting to many thousands of species of plants and animals, was considerably more diverse than it is now. Although it is almost embedded in our culture that much of the present-day New Zealand biota has descended from these Gondwanan founders, there are actually very few examples confirmed by modern molecular studies. What we do have is a growing number of molecular studies that indicate that many of our plants, particularly the herbaceous alpine ones, have reached New Zealand's shores only within the last ten million years or so. Does this necessarily mean 'goodbye Gondwana'? … Or does it simply mean that our initial molecular enthusiasm has been directed overwhelmingly at the recent arrivals? …

There is no question that people have strong views on which of the present biota have descended from the founders and which have colonised New Zealand since then. … Some people lean toward the 'Moa's Ark' end of the spectrum, in which New Zealand is seen as a museum full of ancient Gondwanan relicts. This view is based on a few iconic examples such as tuatara and the native frogs, for which there seems no alternative explanation. At the other extreme are those who advocate a total wipe-out of all previous New Zealand life as a result of the so-called Oligocene drowning event two-thirds of the way through our history … and hence that it is distinctly possible that New Zealand's life began on an isolated archipelago of islands 25 million years ago. This implies that the entire terrestrial fauna and flora of the early Cenozoic have been lost, and that today's organisms have developed from new founders that dispersed across the sea since then. If there is a pattern, it is that many zoologists tend to gravitate towards the first view, while a few botanists and geologists prefer the second, because their respective evidence comes from different sources and plants and animals probably have rather different histories. …

Endemism, or rather the degree of endemism, has been a powerful factor in the Gondwanan heritage argument. Most biologists would accept all of the endemic families of animals that occur in New Zealand … as candidates for the founder biota, even without a fossil record or molecular evidence, because it takes a very long time (deep time) for an organism to diverge sufficiently from its relatives to be ranked as a distinct family. Thus we might expect vertebrates such as the giant ratites (moa and kiwi), the tiny New Zealand wrens, the wattle birds, the New Zealand frogs and tuatara to qualify as candidates for foundation membership.

In the case of the frogs and the wrens this deep-time expectation is confirmed by molecular evidence, but with the others the work has yet to be done (wattle birds) or the molecular evidence does not fully support the Gondwanan vicariance argument (ratites). With ratites, the dates proposed for moa divergence seem to fit the Gondwana model (although there is some circular reasoning here), but it appears that our iconic little kiwi did not diverge from its nearest living ratite relatives until after New Zealand had become an island. We are left with either an island-hopping kiwi ancestor or a lack of faith in the precision of this molecular methodology. The Gondwanan antiquity of tuatara will have to remain in the unconfirmed category until some new technology is able to shed light on its age, because there are no extant close overseas relatives with which to compare the DNA.

Antiquity, by itself, is no absolute guarantee of continuous existence in New Zealand until the present day … we still have the Oligocene marine transgression to contend with. If this was as severe as some are suggesting, it would have extinguished all terrestrial life. If the animal or plant that was lost during the drowning also occurred nearby, then there is the possibility that it might have been reintroduced to New Zealand once the land rose from the sea again. For the vertebrate groups cited above, there is no evidence of their existence in Australia (or anywhere else for that matter) in the past 25 million years, so it is difficult to see how they could have been reintroduced to New Zealand. The endemic bat family, Mystacinidae, is an exception because bats can fly and because three species of fossil bats (*Icarops*) in this family are known from northern Australia during the early to mid-Miocene, dated at about 20 million years ago. Furthermore, the molecular evidence of separation from their nearest relatives in South America gives a figure of 50 to 60 million years ago, which is after the separation of Zealandia. So the conclusion is that these bats dispersed to New Zealand but have since become extinct in Australia. It is quite possible they have been here only since the Oligocene but it is equally possible that they pre-date that era and survived the drowning. With the five endemic families of invertebrates, the necessary level of molecular work has not been done to determine where their nearest overseas relatives might be living or when they diverged. They are really in the 'unknown' category when it comes to the Gondwana question.

So, when we subject the 14 family-level endemic groups of animals which have the potential to be the basis of our Gondwanan heritage to close scrutiny, we find that only the frogs, moa and wrens are backed up by DNA analysis at this stage. Tuatara may easily be included, but the remainder are yet to be tested. Clearly we must go well beyond these family-level organisms if we are to find convincing evidence for our Gondwanan ghosts.

The brevity of the above list of animals certainly does not imply that these were the sole occupants of Zealandia to leave descendants today. Gondwanan invertebrates would obviously have been abundant in the early days of Zealandia, but in the absence of any fossil record, and with very limited molecular work to draw upon, we are left with little to argue from if we want scientific verification. Nevertheless, the insects and other invertebrates still provide, in my view, the strongest of all arguments for the survival of Gondwanan founders throughout New Zealand's history. In particular, the entire community of freshwater invertebrates can be cited as potential Gondwanan descendants … These little animals—ranging from mayflies to sandflies ... and including many non-insect invertebrates such as koura (freshwater crayfish) and kakahi (freshwater mussel)—are totally dependent on their cool freshwater habitat. Sea water is poison to them. It is difficult to see how they could reach New Zealand without having been part of its landscape throughout its island history. Most have not even reached the Chatham Islands from the New Zealand mainland, despite having four million years of opportunities.

Peripatus (the velvet worms) and our large carnivorous snails are frequently cited as prime examples of our Gondwanan fauna. Again the evidence is not as convincing as it could be. A modern DNA analysis of the peripatus group has concluded that the New Zealand species are more closely related to those in Tasmania than to those on mainland Australia. ...

New Zealand's giant carnivorous land snails ... appear to be incapable of oceanic travel. The world distribution of their family (Rhytidae) includes parts of Gondwana—South Africa, Madagascar, eastern Australia and New Caledonia—suggesting they could be a Gondwanan founder in this country. But they also occur on a number of oceanic islands from the Seychelles to Indonesia and Melanesia. The inclusion of oceanic islands implies a capacity for trans-oceanic dispersal, so until further analyses of their history are completed we cannot be totally confident of their Gondwanan heritage.

What about the plants? The vegetation at the time of separation would have looked very different from that of today, but the pollen record shows that many ferns (spore-bearing plants) and seed plants (conifers and flowering plants) that feature in the flora today were here at the time of separation. ...

The magnificent araucarian conifers, even more impressive trees than their podocarp relatives, are represented in New Zealand today only by the kauri (*Agathis*), although there is macrofossil material of *Araucaria* in the Miocene. These great trees had their heyday in the Jurassic and Cretaceous when they were distributed across both hemispheres. However, towards the close of the Cretaceous their stronghold was in Australis. Fossil pollen of *Araucariacites* (fossil *Araucaria*) was present on the pre-New Zealand land area as long as 230 million years ago during the Triassic.

Among the flowering plants, the southern beeches (*Nothofagus*) are the most frequently cited example of a plant group with a Gondwanan history ... These beeches are distributed from Patagonia to eastern Australia, with outliers in New Caledonia and New Guinea, and have a deep fossil history. Their pollen record is excellent, extending beyond their present distribution by including south-western Australia and Antarctica, and it goes back over 80 million years. It enables palynologists to clearly document the trees' waxing and waning in association with climatic changes and their persistence across the time gulf from the Upper Cretaceous to the present. But it seems the more we learn about southern beeches, the more complicated their story becomes ... a molecular study of the modern *Nothofagus* species in New Zealand has shown that their presence can be explained only by long-distance dispersal. This indicates that the continuous pollen record that we find in New Zealand since the early Eocene is likely to represent overlapping extinctions and re-invasions of beeches—events that are not detectable from the fossil record alone. The whole picture has become much more complex than the classic vicariance explanation, and could have implications for many other common plants. ...

From an expected flora of more than 3,000 species of vascular plants at the time of separation from Gondwana, it is clear that we know very little about its make-up. The vast majority are extinct today. It seems that the list of 'Moa's Ark' plant species is shrinking as each new molecular analysis is completed. Nevertheless, there is no reason to abandon the Gondwanan foundation hypothesis just yet. The very fact that some of today's plant groups, like kauri or rimu or *Gunnera*, and probably many more of the animals, could survive here for 80 million years suggests they were not alone. At the time of writing, molecular studies are swinging the vicariance/dispersal debate strongly towards dispersal. This could well be a matter of sampling bias, rather than a fundamental shaking of the composite nature of our biota. One thing we do know is that New Caledonia and New Zealand are repositories for at least some of the founding plant types. ...

I have long been interested in a rather special family of tiny, primitive moths known as the Micropterigidae. Recently, with the collaboration of a team of molecular biologists in Japan led by Yukimasa Kobayashi, we have sequenced a gene of micropterigid species from all over the world and come up with a world phylogeny for this family and some tentative dates for the separation of its major lineages. There is no question that these moths are very ancient; in fact, they could be living representatives of the original moths. Beautifully preserved fossil micropterigid moths have been found in Lebanese amber, dated about 120 million years ago in the mid Cretaceous. ... So we can say with confidence that micropterigids were certainly around long before Zealandia set sail; they looked exactly the same as the moths we find today and, moreover, they were probably worldwide. From the molecular data, it appears that separation of the ancestors of a large New Zealand/New Caledonian contingent from a widely spread Southern Hemisphere lineage appears to have occurred about 80 to 60 million years ago. This would have placed them on the right piece of land at the right time to become part of the biota of Zealandia. Again, from what we know of their sedentary forest lifestyles, there is little chance these tiny ghosts of Gondwana would be here today unless they had both 'caught the boat' and survived the Oligocene. Apart from their distribution on all major continents, they occur only on continental islands like Madagascar, New Caledonia and New Zealand, and there is no evidence of dispersal in their world phylogeny—strong support for the view that they travelled along with the continents.

The conclusion is that descendants of some of the 'ghosts' certainly live in New Zealand today. Exactly how many present-day New Zealand organisms have this ancestry will probably never be known, nor does it need to be. The proportion of the biota that can claim such ancestry will vary greatly between different types of animals and plants. Popular accounts tend to focus uncritically on the same examples time and again, but the vast majority of New Zealand's organisms are the lesser-known insects, other arthropods, snails and worms that have not been investigated. Who knows how many of these belong in the vicariant category and how many came by dispersal? We have examples of both, and both are clearly possible. The proportion of the biota that they make up is not the big issue, as Joseph Hooker observed so long ago when he pondered the botanical connections between the southern lands and islands in 1853, saying 'by a few Chilean plants the whole flora of New Zealand is connected with that of South America.'

From *Ghosts of Gondwana: The History of Life in New Zealand* by George Gibbs: Craig Potton Publishing, Nelson, 2006.

Mismatch

Auckland medical researcher Peter Gluckman (born 1949) is a world leader in the biology of the foetus and the newborn child. His main areas of research are intrauterine growth retardation and brain injury resulting from oxygen deprivation at birth, but his work at the University of 'Auckland's Liggins Institute, which he directs, encompasses the complex links between genetics, environment and disease.

With his British colleague Mark Hanson (born 1949), an honorary visiting professor at the Liggins Institute, Gluckman has written two books, *The Fetal Matrix* and *Mismatch*, from which this passage is taken. As he recounts, it was his work in the Himalayas in 1972 that set the course for his life's work.

'How on earth can anyone bloody well live here? All I want to do is collapse and die!'—this is almost inevitably your first thought if you climb or trek to an altitude above 3,500 metres. Unless you are well-prepared by having taken the time to become acclimatised, all you can do is take two or three steps, then pause for some gasping breaths before taking another two or three steps, then more gasping and another two or three steps—it never seems to end. You are totally unsuited—*mismatched*—to being at such a high altitude. Yet the reality is that people do live at such high altitude, and even higher, in the Himalayas and in the Andes, and have done so for many generations.

The Sherpa are one such people. They are of Tibetan origin but crossed the Himalayas to live in the high valleys of Nepal hundreds of years ago. Until they were 'discovered' by the great Himalayan climbing expeditions of Mallory, Hunt and others in the middle of the twentieth century, they lived in almost total isolation. Essentially their only contacts came through trading expeditions, particularly to obtain rock salt, into Tibet. The Sherpa are Buddhist subsistence farmers living on potatoes, barley and yaks. In more recent years porterage to support climbing expeditions and trekking tourism has changed their economy, but at a considerable cost to their traditional society and environment. The influx of tourists has caused deforestation because the slowly growing mountain trees have been used for firewood. For some reason the tourists do not like their food cooked the traditional Sherpa way over dried yak dung—although frankly the food really tastes much the same however it is cooked over a smoky fire. The point is that the physical and social environment of the Sherpa has changed rapidly.

But in 1972 tourism had not yet arrived in the upper Khumbu valley which leads up to Mount Everest, and down which the Dukh Khosi river tumbles past the famous Thangboche monastery and Namche Bazaar, the biggest Sherpa village. Thirty years ago, this valley was really only known to the climbing community and to a few people working with Sir Edmund Hillary, the New Zealander who with Sherpa Tenzing Norgay was the first to conquer Everest in 1953. Hillary then chose to devote much of his life to supporting the Sherpa people through the development of schools and basic infrastructure, including a small hospital, bridges, an airstrip, and engineering works to protect some of the monasteries. One of us (Peter) travelled there as a newly graduated doctor to assist a medical research expedition initiated by Hillary. The aim of the expedition was to study major health problems resulting from iodine deficiency—these were the price the Sherpa unwittingly paid for living in this extreme environment with its steep terrain and high snowfall. While these lofty valleys might be free from the intertribal conflict and inadequate pastures which had led their ancestors to migrate over the mountains several hundred years earlier, the environment posed other major problems. ...

Over millions of years the Himalayan mountain range, which is very young in geological terms, has been formed by the collision of the Indian and Asian tectonic plates. The Indian plate pushed under the Asian plate as it moved inexorably northwards, driven by convection in the Earth's mantle, generating earthquakes and pushing up the mountains. As the Himalayas grew they were subject to repeated deluges of rain and snow. This washed the soil almost completely free of some minerals, in particular iodine. The result is that the Himalayan foothills are perhaps the most iodine-deficient region of the world. All humans require some iodine in their diet, but this was not possible for the Sherpa in their mountain environment. Even in the mid twentieth century, they were almost totally isolated from access to western foods supplemented with iodine.

The expedition that Peter joined found that over 90 percent of the Sherpa population had goitre. This is the medical term for a grossly enlarged thyroid. The thyroid is a gland located in the neck just below the voice-box which manufactures and secretes a hormone called thyroxine (or thyroid hormone) into the bloodstream. This hormone is made from a combination of an amino acid and iodine, the iodine being absorbed from the diet by the gut. Thyroxine is essential for normal function because it determines the body's rate of metabolism—acting in some ways like the accelerator of a car. If the accelerator is pushed down too far (by having too much thyroxine) the engine (the body's metabolism) revs too fast. If there is not enough thyroxine, the body's metabolism slows down and does not have enough power to function properly. So the result of the low iodine in the diet was that many Sherpa had slow metabolism. The goitres that Peter saw in the Sherpa were sometimes larger than the neck itself, giving a grotesque appearance. Unfortunately, enlarging the gland does not solve the problem—even a bigger gland cannot make thyroxine without adequate iodine in the diet. And so the inevitable signs of slowed metabolism appeared, such as delayed reflexes, fluid retention, higher blood-fat levels, and poor heat generation by the body.

The secretion of thyroid hormone is controlled by a very elegant system involving another hormone, thyrotropin, made by the pituitary gland, which is the so-called 'master gland' at the base of the brain. If the thyroid hormone levels in the blood are low, more thyrotropin is secreted and the thyroid gland is driven to secrete more thyroid hormone. Conversely if thyroid hormone levels are high in the blood, less thyrotropin is secreted.

This kind of control system is called a negative-feedback loop and is a common way of maintaining constancy within a biological system. It is analogous to a thermostat-controlled heater— as a room gets colder, the thermostat signals to the heater that it must generate heat; when the temperature has risen to the preset level the thermostat turns off, but it clicks on again if the room starts to cool. This is not an entirely closed system because you can change the setting on the thermostat, and in just the same way biological systems acting through the brain's control of the pituitary gland can alter the requirements for thyroid hormone. But if thyroid hormone levels remain low for a long time, for example because the diet does not contain sufficient iodine, the continuously high levels of thyrotropin also stimulate growth of the thyroid gland in a desperate attempt to make more thyroxine. As the gland enlarges, it becomes visible in the neck as a goitre.

Of even greater concern to the doctors on the expedition was that one in eight of the population showed a particularly tragic consequence of iodine deficiency, one that started before birth. In some people the lack of iodine during foetal life drastically affected their brain development. They were born as cretins. This is a medical term (unkindly used in a pejorative way by people who have not witnessed the condition) for the severe mental retardation associated with intrauterine iodine deficiency. Yet these *Kurs* (the Sherpa word for these individuals) were generally well-integrated into society and had valuable jobs. They were the water carriers—spending their days carrying buckets of water from streams in the valleys to houses on the high terraces.

Despite much that was known about cretinism at the time of the expedition, there was still a major scientific mystery. Not all Sherpa, despite their very low iodine intakes, showed these signs of iodine deficiency. And even more striking was the fact that not all babies were born cretins, even though virtually all their mothers were iodine-deficient. On closer examination the mystery deepened still further. Some cretins had a particular form of cerebral palsy which resulted from their iodine deficiency interfering with brain development in an irreversible way. But others did not. Some cretins were extremely dwarfed (less than 1.4 metres tall when fully grown) but others were not. And some were deaf-mute whilst others were not. All these various clinical pictures appeared to be different manifestations of the same environmental deficiency: lack of iodine. And when Peter and his fellow doctors treated potential mothers in this population with iodine injections, all the different forms of cretinism disappeared, showing the central role of iodine deficiency in producing them.

Here were some important lessons for a young medical scientist. The first was that not all individuals show the same symptoms and signs, even when faced with the same conditions. Clearly the potential for goitre, thyroid hormone deficiency and cretinism was the result of an interaction between some intrinsic susceptibility, perhaps based on individual variations in genetic make-up, and the environment in which they lived. The source of such variations was not always clear. It might be the individuals who varied, or their environment, or both. For example, one village, Phortse, had a particularly high rate of cretinism. Phortse was located away from the other villages and made a beautiful picture after the snow had fallen, with its two-storeyed stone houses—yaks living below, people living above—scattered across ancient stone terraces on the side of the valley. Was the particular problem in Phortse due to some local genetic variation predominant in the families that lived there? Or was it due to their diets being slightly different from the other villages? On investigation, the only difference seemed to be that the villagers of Phortse ate much more barley. While the doctors could not prove it, they thought that the barley might contain something that interfered with thyroid gland function. This was not implausible because in Tasmania children had been shown to develop goitre in the spring, but it then disappeared later in the year, only to return the following year. The effect turned out to be due to contamination by a chemical in cow's milk which came from the wild turnip that appeared in the pasture every spring.

So the messages were clear—the environment of a population can vary in subtle ways which may not be immediately obvious, but which nonetheless can have dramatic effects on the pattern of disease in that population. Not everyone has the same constitution and so not everyone responds to any particular environment in the same way. The more we look, the more we realise that this principle applies not just in Nepal, but across the world. And even subtle changes in the environment can have a major impact, depending on the nature of the change and when in the person's life course it occurs. This variation in our constitution and our ability to match our biology to our environment is central to understanding how we live in this world, and whether we remain healthy or develop disease.

The Sherpa were paying a severe price in terms of their health for living in high Himalayan valleys. But in many other ways as a population they had adapted astonishingly well to this rugged place, and that is why they stayed there. They had overcome the problems of breathlessness (unlike Peter!) and could carry enormous loads to high altitudes. While the normal porter's load was 30 kilograms, they were able to carry a double load—and be paid twice as much. One of the Sherpa porters on the expedition would regularly carry a load of over 60 kilograms—a good deal more than he weighed himself. He had clearly developed strong muscles as well as lungs.

But it was not just in terms of their anatomy and physiology that the Sherpa had adapted to their environment. They had also developed over many generations a complex social structure and a sophisticated culture. This culture differs in many ways from that to which most of us are accustomed. For example, some practised polyandry, the practice of a wife having several husbands. Even more surprisingly, often her multiple husbands were brothers. Such customs appear to assist the Sherpa in coping with their extreme environment, in this case because the wife needed to have a strong man around at home even if her other husbands were off tending the yak herd.

For Peter's expedition, the urgent question was what could be done to deal with the problem of iodine deficiency in the Sherpa. This was not easy to answer. In Europe the solution would be to add iodine to a foodstuff, as had been done with the introduction of iodised powdered table salt. Such measures, taken in the 1920s, quickly prevented goitre in parts of the UK. The condition of 'Derbyshire neck' was no longer seen. The 'fashionable' distended throats of young women who sat as models of doomed heroines for the Pre-Raphaelite painters of the middle of the nineteenth century became a thing of the past. But while powdered salt might have seemed the appropriate way to bring iodine to the Sherpa, the researchers found that Sherpa tradition demanded that they use rock salt brought over the high altitude passes (about 5,800 metres) from strife-torn Tibet. There were several strong cultural reasons which could be identified for continuing such trade: it preserved communication between groups of the same religion; and it was part of the expectation of the young males that they should face the dangerous passage across the icy storm-weathered high altitude passes. Such perspectives had to be understood in attempting to find a solution to the problem. The doctors ended up by injecting each Sherpa with a depot of iodine, a treatment costing only about ten cents, which lasts at least five years. It is a solution that has been adopted in several other re-mote mountainous areas such as New Guinea where food-based alternatives such as iodised salt cannot be reliably applied.

When we think about such lessons, drawn from working and living with the Sherpa in the Himalayas, some more general biological insights appear … The first is about ability to adapt (adaptedness). Clearly humans can live in some very extreme environments, well away from the savannah of central Africa where we first evolved. We can become matched to a variety of environments because, like some other animals such as the rat and cockroach, we have broad adaptive capacity. In contrast, many other species of animal are exquisitely matched to a par-ticular environmental niche. No one can doubt how well the emperor penguin is adapted for a life fishing in frigid ocean waters and breeding on the Antarctic ice shelf; or how well the cheetah is suited for sprinting at over 100 kilometres an hour to run down an impala on the savannah; or indeed how well the chameleon or the stick insect have perfected the art of camou-flage to make them well hidden from predators in their habitat.

The second point relates to cost. The polar bear is not adapted for life in the tropics, nor the Malayan sun bear for life in the Arctic. If they were to be transported from one to the other, they would perish. Similarly, humans have evolved to be able to live in a broad range of environments, but nonetheless we are not infinitely adaptable. Although we can often cope at least for a while when we get to the extremes of this range, if we try to live beyond these environmental limits there will be a cost. So a species thrives if it lives in an environment for which its 'design' matches it. The greater the degree of mismatch between environment and design, the greater the cost ... we call this the 'mismatch paradigm', and the cost of such mismatch is often disease, just as the cost to the Sherpa of inhabiting the high Himalayan valleys was goitre and cretinism. As we carried out our research, we wondered whether even the so-called 'modern environment' in which so many of us aspire to live lies beyond the limits of human adaptedness and, if so, how? Could there be a cost even for contemporary societies living in developed countries? ...

Humans have chosen to live in an enormous range of environments. The Sherpa live at high altitude, the Inuit above the Arctic Circle, the Fuegians—described in rather uncomplimentary terms by Darwin in his book *The Voyage of the Beagle* —lived at an extreme southerly latitude, the Tuareg live in the Sahara desert. These are all examples of human existence in extreme natural environments. Other populations live in threatening environments created by humans themselves. Parts of southern California and the Australian outback are becoming more difficult to live in as the underground aquifers are drained to provide water for cities and industry, and the increasingly salty soil cannot support plant growth. Nauru Island in the Pacific sustained a stable society for over a thousand years until colonial powers removed all the topsoil to mine the nitrate-rich guano in their voracious desire for fertiliser—from being a rich, vibrant society on a luscious tropical island, Nauru is now a horrific landscape of bedrock and there is discussion about relocating the remaining islanders to Australia or New Zealand. The Easter Islanders lived (until they nearly all died out) with the consequences of tree destruction on their remote Pacific island—they could no longer build boats and so fish could not be part of their diet. Other Polynesians had to live with the consequences of overcrowding on small islands and some populations went as far as widespread infanticide to control their numbers. The Japanese fishermen of Minamata Bay had to live with the consequences of mercury poisoning which caused tragic brain damage to their children. Some soldiers who served in Vietnam may live with their exposure to Agent Orange and it is possible that these effects are transmitted to their offspring. The list of tragedies goes on and on. They are all examples of how, even without migrating to remote places, we really can make our environment challenging to inhabit.

The most serious consequences were seen in the Sherpa when the iodine deficiency occurred during foetal life, as this led to cretinism. Similarly in Minamata Bay and the Vietnam veterans it was the impact of the environmental crisis on the developing foetus or infant which had the most dramatic consequences. ...

The developmental perspective has had surprisingly little influence in forming our understanding of the human condition. Yet embryology was a very important component of the nineteenth-century biologists' research, including that of Darwin himself. He recognised that the complexities of embryonic development might reveal much about how different species evolved and related to each other in the evolutionary tree. But the enthusiasm for embryology was lost in the early twentieth century when biology became dominated by the growing understanding of genetics. It is only recently that we have started to appreciate again how important an understanding of development is to the whole of biology.

We now understand that environmental exposures during development are associated with choices that both predict and determine the environments to which we are best matched. Our biological processes are designed to respond to signals coming from the environment, and to induce responses which can either be for immediate biological advantage (such as burning brown fat, a form of energy reserve in newborn babies, to generate heat when they are cold) or which help survival in the future. A grizzly bear puts on a lot of fat in the autumn so that it can use this stored energy to support body functions while it hibernates. The laying down of fat has no immediate advantage but is done because of the biological expectation of an impending winter. ... Animals do not gaze into crystal balls to foretell the future, but evolution has equipped them with the ability to gain information from their environment and use it to adjust their biology for future advantage. The grizzly bear's metabolic biology has evolved to be sensitive to shortening day length. And this type of biological forecasting even starts during foetal development.

So the developing organism uses information from its environment to make choices in an attempt to match its constitution to the environment it forecasts it will inhabit. In the same way, when we pack our bags for a journey we try to predict the weather we will face, and choose our clothes accordingly. If we have limited luggage, for example because we are travelling by air, we make choices about what to take and what to leave behind. Perhaps if we take an umbrella we will not take a raincoat. If we are going skiing, we will leave behind our shorts and sandals. Similarly, the embryo and foetus try to forecast their future environment and choose their luggage—in other words, they choose what kind of adaptive preparations and strategy will maximise their chance of reproductive success, because this is the ultimate goal of their biological journey. If they predict being born into a cold environment, they may develop a thicker coat of fur. This is what happens in the Pennsylvanian meadow vole, a small creature which looks like a cross between a mouse and a hamster. Like all voles, the meadow vole grows fast and mates within a few weeks of being born. If they are born in spring the pups have a thin coat of fur, but if they are born in autumn they have a much thicker coat. This difference in coat thickness is permanent and clearly has a survival advantage because winters of the north-eastern United States are frequently harsh even though the summers can be warm. More dramatically, if an organism predicts that it will be living in an environment where there will be very many hungry predators, it may invest some of its developmental resources in defence. *Daphnia*, also known as the water flea because of the jerky way in which it swims, is a small freshwater crustacean that is popular with aquarium enthusiasts as fish food. In natural ponds, insect larvae are one of the main predators of *Daphnia*. But these larvae release chemical clues (called kairomones) which give away their presence in the pond, and if growing *Daphnia* detect high concentrations of these chemicals then they develop with a sort of body armour that makes them less vulnerable as prey. ...

So how have we become adapted to the environments we inhabit? In part this is by natural selection, a term first introduced by Charles Darwin in his most important book, *On the Origin of Species by Means of Natural Selection* (1859). This is the process by which organisms evolve through the selection of variations in characteristics, sometimes called 'traits', which make them more likely to survive and reproduce. Genes that influence the expression of biologically advantageous characteristics in individuals in a particular environment are more likely to be passed on to the next generation, because individuals possessing them are more likely to reproduce. So the mix of genes within a population—that is, the amount of genetic variation in the 'gene pool'—is changed over time and the characteristics of the species are gradually refined to match the environment.

But within a life course the environment also influences how the genes each individual has inherited from its mother and father are turned on and off. In the so-called 'plastic' phases of embryonic, foetal and infant life, environmental influences can mould how our characteristics develop, with permanent consequences. Evolution has equipped us with particular ways of responding to the environment during development and as a result there is a range of developmental choices and trade-offs which we can make; these are used to improve the chance of a match with the environment. Depending on the circumstances, the end result may be good or bad.

A match means two things are complementary. The shoes you are wearing match but are not identical, unless you have two left feet! When we say that two people in a relationship are well matched, we don't mean that they are very similar. In fact we may mean the opposite, that they understand one another well and that any shortcomings in the personality or behaviour of one are complemented by the other. When one is fed up or tired, the other will take action to help. The relationship works because it is dynamic, each person responding to the needs of the other in a mutually supportive way. We are concerned with such a complementary relationship, that between the biology of humans on one hand and the nature of the environments in which we live on the other. Each is continuously changing to an extent and there is a constant dialogue between them. If the organism is *matched* to its environment, then we suppose that it has become suited by both evolutionary and developmental processes to be so.

A match does not have to be 'all or nothing'—there are degrees of match and degrees of mismatch, just as we might look in the mirror and ask ourselves how well a jacket matches the shirt we are wearing. The greater the degree of match between an organism's constitution and its environment, the more likely the organism is to thrive; the greater the degree of mismatch, the more the organism has to adapt or cope. This incurs costs, and these rise as the degree of mismatch increases. If the organism cannot cope at all, then the consequences will be a greater risk of disease or death. Mismatch can be created by changes in either the organism or its environment. It might arise, for example, as a result of a mutation in a gene essential for the organism's healthy life in that environment. The disease of lactose intolerance results from a mutation in the gene responsible for synthesising the enzyme lactase. In the gut wall this enzyme breaks down the sugar lactose, found in foods such as cow's milk, making it possible to absorb it. If a person with lactose intolerance lives in an environment where cow's milk is a staple food, they will suffer from chronic diarrhoea from unabsorbed sugar remaining in the gut and they will also become malnourished.

Alternatively, mismatch can arise from the environment changing rapidly or drastically. Sailors in the eighteenth century had a high chance of dying on long voyages from scurvy, a very unpleasant condition in which bleeding occurs into many of the tissues of the body. The skin shows widespread bruising, gums bleed, teeth loosen, and recently healed wounds may break down. Bleeding into muscles and joints causes considerable pain. Scurvy developed because at sea the sailors' nutritional environment had changed from one where there was some fresh food to one in which all food had to be dried or salted. There were no fresh vegetables or fruit until they reached the next port, which might be months away, so their diet was very deficient in vitamin C. The human body stores some vitamin C and it takes time to run these stores down, but when they run out scurvy results. This is why the disease only became apparent on long voyages. But consuming only a small amount of fresh fruit could help. The sailors must have argued about this issue: how could fruit be kept for months in the warm, wet hold of a ship? The answer was to take vegetable extracts and then later lime and lemon juice on the voyage. The sailors did not like the taste, preferring rum, but sauerkraut and limes saved the day, Britannia ruled the waves and Britons abroad became known as 'limeys'. ...

Many humans now live in environments to which they are not well matched, and for many the degree of mismatch is increasing. Mismatch occurs in everyday existence in seemingly ordinary environments. And mismatch has a cost—it turns out to be a major determinant of our social structure, health, and quality of life. Understanding the mismatch paradigm gives us a new perspective on what we are.

From *Mismatch: Why Our World No Longer Fits Our Bodies* by Peter Gluckman and Mark Hanson: Oxford University Press, Oxford, 2006.

To Find a Planet...

In January 2006 two University of Canterbury astronomers, husband and wife Michael Albrow and Karen
Pollard, were part of a team that made international news
headlines with their discovery of an extra-solar planet
that was the most Earth-like planet yet found. As Karen
Pollard (born 1966) relates here, the quest to find this
planet, now known as OGLE-2005-BLG-390Lb, took ten
years.

We began our planet-hunting in 1995—by coincidence, the year
the first-ever discovery of an extra-solar planet was announced to
the world, in *Nature.* Early that year, astronomers Penny Sackett
and Kailash Sahu had sent an email around the world asking
astronomers if they would like to participate in an international
collaboration to search for planets, using a new technique called
microlensing. At that time, Michael and I were on post-doctoral
fellowships at the South African Astronomical Observatory in
Cape Town; we had arrived in January, just a few months after
Nelson Mandela had come to power as president and leader of
the ruling African National Congress.

It was a fascinating time. South Africans seemed very positive
about the changes happening in their country and were working together for a better future, introducing innovations such as
free (and compulsory) education for all children, free medical
care and new housing schemes. SAAO was a very active research
centre and a great place to work. Although our offices were in
Cape Town, we had excellent access to good telescopes and
instruments at the observatory in Sutherland, which was in the
desert-like Karoo area of Northern Cape Province—a four-hour
drive away.

Michael and I thought that Sackett and Sahu's microlensing project sounded novel and exciting. The gravitational microlensing method uses the gravity of a dim foreground star (a 'lens star') to act as a giant natural lens, magnifying a more distant, bright star (the 'source star'). If a dim star somewhere in our galaxy passes slowly in front of a brighter background star, the gravitational influence of that foreground star can bend and magnify the light of the background star. While we never actually *see* the foreground star, we see the background star getting brighter and then fainter as the foreground star passes in front of it.

Einstein predicted this effect in 1936, but because of the very precise stellar alignment required he never expected it to be observed. In the early 1990s, however, several massive observational campaigns—MACHO, OGLE and EROS—were launched to survey millions of stars, and to use the microlensing effect to search for elusive 'dark matter' in the spherical halo that surrounds the disc of our Milky Way galaxy.

During the first international meeting in California in late 1994, the major microlensing collaborations had pledged their intention to provide public alerts to microlensing events in real time. Sackett and Sahu realised that if a lens star had a planet orbiting it, the light output from the microlensing event would be distorted by the gravitational influence of the orbiting planet. However, to detect this small anomaly—which could last days for a giant planet or only a few hours for a planet with a similar mass to Earth's—astronomers would need round-the-clock monitoring of any microlensing events. This would happen only with a worldwide network of telescopes, so that it would always be night-time at one of the telescopes, and constant observations could be undertaken —provided the weather was clear. To achieve this, Sackett and Sahu had sent their email to various observatories and research institutes around the world, and so in mid 1995 we became founding members of their PLANET (Probing Lensing Anomalies NETwork) collaboration, and started observing.

PLANET's observations began with four southern hemisphere telescopes located in South Africa, Chile and Australia. (Since then, to enhance our ability to observe around the clock, we have added other telescopes at a variety of longitudes.) We relied on microlensing alerts from MACHO and OGLE, which we would then monitor very precisely to detect and study any anomalies that could be evidence of an orbiting planet. Further micro-lensing detection and follow-up collaborations were started over the next ten years, including MOA, a New Zealand–Japan joint venture that now operates a telescope at the University of Canterbury's Mount John Observatory near Tekapo in the South Island, and microFUN, an informal microlensing consortium that includes some other New Zealand astronomers.

In 1997, Michael and I left South Africa and returned to the department of physics and astronomy at the University of Canterbury, where, apart from a year in the United States, we have remained. We continued our work with PLANET by helping to plan and coordinate the observations, by travelling to South Africa and Hobart to make observations, and by writing some of the computer software for reducing and analysing the data we were getting from the telescopes.

On July 11, 2005, the OGLE search team discovered the micro-lensing event OGLE-2005-BLG-390—the 390th event they had detected that year. This triggered the PLANET telescopes to start taking data from an event occurring close to the centre of the Milky Way, some 25,000 light years from Earth. A light curve consistent with a single lens star had its maximum brightness on July 31.

Then, on August 10, PLANET member Pascal Fouqué, observ-ing with the Danish telescope at La Silla, Chile, noticed in the data a planetary deviation—a 'bump' on the light curve caused by the presence of an orbiting planet. Because our images of the sky are processed immediately by computer, we get instant feedback on the events we are looking at. As soon as the bump was noticed, an alert was sent to the PLANET collaboration coordinator or 'home-base', the person in charge of prioritising which events to observe, and letting everyone know when a possible planetary anomaly is under way. Everyone who was observing kept in constant touch by email or telephone, and those of us who weren't observing were logged on to a PLANET website, watching the light curve change with each new observation that came in. Anomalies are always exciting, even when they don't turn out to be planets.

An observation from the same night's OGLE team also showed the anomaly, while the last half of the planetary deviation, which lasted about a day, was covered by images from Andrew Williams at Australia's Perth Observatory. The MOA collaboration also confirmed the deviation with their observations.

The final analysis of the light curve, using data from the three collaborations, revealed the smallest and most Earth-like planet yet discovered. According to our calculations, this planet has a mass five and a half times that of Earth, equating to a diameter almost twice that of Earth. It orbits about two and a half times further away from its parent star than we do from our sun: if it were in our solar system, its orbit would be somewhere between those of Mars and Jupiter. And its parent star is about one-quarter the mass of our sun.

The combination of the greater distance and the fainter, dimmer parent star means the planet is likely to be much cooler than Earth; we have computed its surface temperature as minus 220°C. But having a solid and frozen surface, rather than being a gaseous ball, is what makes it the most Earth-like planet discovered. One smaller planet has since been found by the radial velocity technique, in which the presence of a planet is detected though wobbles, or small back and forth motions, of its parent star, but the discovery of a planet as small as Earth in an Earth-like orbit remains elusive.

I took an artist's impression of OGLE-2005-BLG-390Lb in orbit around its dim, red parent star to my children's pre-school and asked the children to think of a name for the planet: although astronomical objects are all officially named by conventions confirmed by the International Astronomical Union, I felt we needed a pet name for our discovery. My daughter and her friends opted for 'Winter'—'because it is so cold'.

Astronomers continue to find planets using this microlensing technique, the radial velocity technique (the most successful so far), and the transit technique, where the planet periodically moves in front of its parent star, dimming it slightly. However, finding planets is very difficult, and our techniques are much less sensitive to smaller planets than to larger ones. Even so, we have found several small planets already, which indicates they could be more common than the larger ones.

We have now found enough planets that we can look at statistics and trends to get better information about how planets and planetary systems form. Knowing how many planets of each mass we have and where they orbit will allow us to estimate how many habitable planets there are in our galaxy, and in the universe as a whole. Who knows, one day we may find a planet with all the important characteristics of our own Earth. I wonder if it will be inhabited?

By May 2008, the PLANET collaboration had contributed to the discovery of four planets. The total number of extra-solar planets found by all techniques—microlensing, radial velocity and transit—was 287.

'To Find a Planet...' by Karen Pollard is first published in this volume.

Sources

Polynesian Navigation
Peter Adds, 'A Brilliant Civilisation', *The Transit of Venus*: Awa Press, Wellington, 2007, pp. 43–53.

The Transit of Venus
2004 Annual Report of the Royal Society of New Zealand:
www.rsnz.org/directory/yearbooks/2004.

General Account of New Zealand
Sir Joseph D. Hooker, ed., *Journal of the Right Hon. Sir Joseph Banks*: Macmillan and Co. Ltd, London, 1896, pp. 222–29.
Additional sources: Peter Adds et al, *The Transit of Venus*: Awa Press, Wellington, 2007; Anne Salmond, *The Trial of the Cannibal Dog: Captain Cook in the South Seas*: Allen Lane, London, 2003.

'Not a Pleasant Place'
Nora Barlow, ed., *Charles Darwin's Diary of the Voyage of H.M.S. 'Beagle'*: Cambridge University Press, Cambridge, 1933, pp. 365–75.
Additional source: George Gibbs, *Ghosts of Gondwana: The History of Life in New Zealand*: Craig Potton Publishing, Nelson, 2006.

Whales and Whaling
Ernst Dieffenbach, *Travels in New Zealand: With Contributions to the Geography, Geology, Botany, and Natural History of That Country*: vol. 1, J. Murray, London, 1843, pp. 44–55.
Additional sources: Jock Phillips, 'Whaling', *Te Ara: The Encyclopedia of New Zealand*: www.TeAra.govt.nz/EarthSeaAndSky/HarvestingTheSea/Whaling/en; Denis McLean, 'Dieffenbach, Johann Karl Ernst 1811–1855', *Dictionary of New Zealand Biography*: www.dnzb.govt.nz.

To Ernst Dieffenbach
Ian Wedde, *The Commonplace Odes*: Auckland University Press, Auckland, 2001, p. 44.

Magnitude Eight Plus
Rodney Grapes, *Magnitude Eight Plus: New Zealand's Biggest Earthquake*: Victoria University Press, Wellington, 2000, pp. 17–25, 51–55, 93–94, 96–98.
Additional source: Eileen McSaveney, 'Historic Earthquakes', *Te Ara: The Encyclopedia of New Zealand*: www.TeAra.govt.nz/EarthSeaAndSky/NaturalHazardsAndDisasters/HistoricEarthquakes/en.

The Remarkable Pink and White Terraces
Ferdinand von Hochstetter, *Geology of New Zealand: Contributions to the Geology of the Provinces of Auckland and Nelson*: C.A. Fleming (trans. & ed.), Government Printer, Wellington, 1959, pp. 153–56, 159–61.

Additional sources: C.A. Fleming, 'Hochstetter, Christian Gottlieb Ferdinand von 1829–1884', *Dictionary of New Zealand Biography*: www.dnzb.govt.nz; D.L. Mundy, *Rotomahana and the Boiling Springs of New Zealand: A Photographic Series of Sixteen Views*: Sampson Low, Marston, Low and Searle, London, 1875.

The Great New Zealand Ice Period
Julius von Haast, *Geology of the Provinces of Canterbury and Westland, New Zealand: A Report Comprising the Results of Official Explorations*: printed at the *Times* office, Christchurch, 1879, pp. 189–91.
Additional sources: Colin J. Burrows, *Julius Haast in the Southern Alps*: Canterbury University Press, Christchurch, 2005; C.A. Fleming, 'Hochstetter, Christian Gottlieb Ferdinand von 1829–1884', *Dictionary of New Zealand Biography*: www.dnzb.govt.nz.

On the Vegetable Food of the Ancient New Zealanders
William Colenso, 'On the Vegetable Food of the Ancient New Zealanders Before Cook's Visit', *Transactions and Proceedings of the New Zealand Institute*: vol. 13, 1880, pp. 5, 11–14.
Additional source: David Mackay, 'Colenso, William 1811–1899', *Dictionary of New Zealand Biography*: www.dnzb.govt.nz.

Beautiful Plumage
Sir Walter Lawry Buller, *A History of the Birds of New Zealand*: 2nd edn, vol. II, published by the author for the subscribers, London, 1888, pp. 7–17.
Additional sources: Ross Galbreath, *Walter Buller: The Reluctant Conservationist*: GP Books, Wellington, 1989; Ross Galbreath, 'Buller, Walter Lawry 1838–1906', *Dictionary of New Zealand Biography*: www.dnzb.govt.nz.

Huia
Anna Jackson, *The Pastoral Kitchen*: Auckland University Press, Auckland, 2001, p. 30.

Search for the Stitchbird
Andreas Reischek, *Yesterdays in Maoriland: New Zealand in the 'Eighties*: H.E.L. Priday (trans. & ed.), Jonathan Cape, London, 1930, pp. 83–93, 98–101.
Additional sources: Michael King, *The Collector*: Hodder & Stoughton, Auckland, 1981; Gerald Hutching, *The Natural World of New Zealand*: Viking, Auckland, 1998; Ray G. Prebble, 'Reischek, Andreas 1845–1902', *Dictionary of New Zealand Biography*: www.dnzb.govt.nz; G.R. Angehr, 'A Bird in the Hand: Andreas Reischek and the Stitchbird', *Notornis* 31, 1984, pp. 300–11.

The Skeleton of the Great Moa in the Canterbury Museum, Christchurch
Allen Curnow, *Early Days Yet: New and Collected Poems 1941–1997*: Auckland University Press, Auckland, 1997, p. 220.

Sources

After the Tarawera Eruption

J.A. Pond and S. Percy Smith, 'Observations on the Eruption of Mount Tarawera, Bay of Plenty, New Zealand, 10th June, 1886', *Transactions and Proceedings of the New Zealand Institute*: vol. 19, 1886, pp. 342–70.

Additional sources: Eileen McSaveney, Carol Stewart and Graham Leonard, 'Historic Volcanic Activity', *Te Ara: The Encyclopedia of New Zealand*: www.TeAra.govt.nz/EarthSeaAndSky/NaturalHazardsAndDisasters/HistoricVolcanicActivity/en; R.F. Keam, *Tarawera: The Volcanic Eruption of 10 June 1886*: R.F. Keam, Auckland, 1988.

The Insect-hunter

G.V. Hudson, *An Elementary Manual of New Zealand Entomology: Being an Introduction to the Study of Our Native Insects*: West, Newman & Co., London, 1892, pp. 9–14.

Additional source: George Gibbs, 'Hudson, George Vernon 1867–1946', *Dictionary of New Zealand Biography*: www.dnzb.govt.nz.

Rutherford and the Discovery of Radioactivity

Ernest Rutherford, 'The History of Radioactivity', in *Background to Modern Science*: Joseph Needham and Walter Pagel (eds), Cambridge University Press, Cambridge, 1938, pp. 55–59.

Additional source: John Campbell, 'Rutherford, Ernest 1871–1937', *Dictionary of New Zealand Biography*: www.dnzb.govt.nz.

A Most Incredible Event

Ernest Marsden, 'Rutherford Memorial Lecture, 1948', reprinted from the *Proceedings of the Physical Society*: series A, vol. 63, 1950, pp. 316–17, Sir Ernest Marsden Papers, MS-Papers 1342-264: Alexander Turnbull Library, National Library of New Zealand; and Ernest Rutherford, 'The Development of the Theory of Atomic Structure', in *Background to Modern Science*: Joseph Needham and Walter Pagel (eds), Cambridge University Press, Cambridge, 1938, p. 68.

Additional sources: John Campbell, *Rutherford: Scientist Supreme*: AAS Publications, Christchurch, 1999; Ross Galbreath, 'Marsden, Ernest 1889–1970', *Dictionary of New Zealand Biography*: www.dnzb.govt.nz; John Gribbin, *Science: A History 1543–2001*: Allen Lane, London, 2002, pp. 504–15.

Booming Kakapo

Richard Henry, *The Habits of the Flightless Birds of New Zealand: With Notes on Other New Zealand Birds*: Government Printer, Wellington, 1903, pp. 18–26.

Additional sources: Gerard Hutching, 'Large forest birds', *Te Ara: The Encyclopedia of New Zealand*: www.TeAra.govt.nz/TheBush/NativeBirdsAndBats/LargeForestBirds/en; Kakapo Recovery Programme: www.kakapo.org.nz.

Kakapo

Sonja Yelich, *Trout 10*: www.trout.auckland.ac.nz/journal/10/10_3.html.

Additional source: Kakapo Recovery Programme: www.kakapo.org.nz.

A Trip to the Seaside

George M. Thomson, *A New Zealand Naturalist's Calendar and Notes by the Wayside*: R.J. Stark and Co., Dunedin, 1909, pp. 191–94.

Additional source: Ross Galbreath, *Scholars and Gentlemen Both: G.M. and Allan Thomson in New Zealand Science and Education*: Royal Society of New Zealand, Wellington, 2002.

The Much-maligned Mangrove

Leonard Cockayne, *New Zealand Plants and Their Story*: 2nd edn, Government Printer, Wellington, 1919, pp. 35–37.

Additional source: A.D. Thomson, 'Cockayne, Leonard 1855–1934', *Dictionary of New Zealand Biography*: www.dnzb.govt.nz.

From Sea Floor to Coastal Plain

Charles Cotton, *Geomorphology of New Zealand*: Dominion Museum, Wellington, 1922, pp. 69–72.

Additional sources: Rodney Grapes, 'Sir Charles Cotton, 1885–1970', in *Eminent Victorians: Great Teachers and Scholars from Victoria's First 100 Years*: Vincent O'Sullivan (ed.), Stout Research Centre, Victoria University of Wellington, Wellington, 2000, pp. 109–25.

Shell Collecting in the 1920s

Charles Fleming, 'Boyhood Recollections of Shell Collecting in the Nineteen Twenties', *Poirieria*: vol. 11, no. 2, 1981, pp. 8–10, Conchology Section, Auckland Institute and Museum, Auckland.

Additional sources: R.K. Dell, 'Fleming, Charles Alexander 1916–1987', *Dictionary of New Zealand Biography*: www.dnzb.govt.nz; Henry Suter, *Manual of the New Zealand Mollusca: Atlas of Plates*: John Mackay, Government Printer, Wellington, 1915; A.G. Beu, 'Suter, Henry 1841–1918', *Dictionary of New Zealand Biography*: www.dnzb.govt.nz.

The Two Lucies

Lucy Moore, 'Auckland Botany in the Cranwell Era', *Auckland Botanical Society Newsletter*, 41, 2 (July 1986), pp. 20–24.

Additional sources: John Morton, 'Moore, Lucy Beatrice 1906–1987', *Dictionary of New Zealand Biography*: www.dnzb.govt.nz; 'Lucy May Cranwell Smith, 1909–2000': www.rsnz.org/directory/yearbooks/year00/cranwell.php; Kirstie Ross, 'The "Two Lucys": The Collaborative Work of Lucy Moore and Lucy Cranwell, 1928–1938', *New Zealand Science Review* 58, 4 (2001), pp. 138–41; L.B. Moore, 'The Structure and Life-history of the Root Parasite *Dactylanthus Taylori* Hook. f.', *New Zealand Journal of Science and Technology*, January 1940, p. 212B.

Tutira: The Story of a New Zealand Sheep Station

H. Guthrie-Smith, *Tutira: The Story of a New Zealand Sheep Station*: 2nd edn, William Blackwood & Sons, Edinburgh, 1926, pp. 292–95, 319–20, 334–35.

Additional source: Ronda Cooper, 'Guthrie-Smith, William Herbert 1862–1940', *Dictionary of New Zealand Biography*: www.dnzb.govt.nz.

Sources

Beginnings: Guthrie-Smith in New Zealand, 1885
Peter Bland, *Stone Tents*: London Magazine Editions, London, 1981, p. 19.

The Maori Way of Birth
Peter H. Buck (Te Rangi Hiroa), *Vikings of the Sunrise*: J.B. Lippincott Company, Philadelphia, 1938, pp. 259–63; Te Rangi Hiroa (Sir Peter Buck), *The Coming of the Maori*: Whitcombe and Tombs Ltd, Wellington, 1949, pp. 350–52.
Additional source: M.P.K. Sorrenson, 'Buck, Peter Henry 1877?–1951', *Dictionary of New Zealand Biography*: www.dnzb.govt.nz.

A Man Who Moved New Zealand
Simon Nathan, *Harold Wellman: A Man Who Moved New Zealand*: Victoria University Press, Wellington, 2005, pp. 74–76, 116, 118–19.

Paradise Lost
Pérrine Moncrieff, 'The Destruction of an Avian Paradise', *Forest and Bird*, May 1949, pp. 3–5; August 1949, pp. 8–11; November 1949, pp. 5–7.
Additional sources: David Young, *Our Islands, Our Selves: A History of Conservation in New Zealand*: University of Otago Press, Dunedin, 2004; Robin Hodge, 'Moncrieff, Pérrine 1893–1979', *Dictionary of New Zealand Biography*: www.dnzb.govt.nz.

The Passing of the Forest: A Lament for the Children of Tané
William Pember Reeves, *The Passing of the Forest and Other Verse*: George Allen & Unwin Ltd, London, 1925, pp. 9–12.
Additional source: Keith Sinclair, 'Reeves, William Pember 1857–1932', *Dictionary of New Zealand Biography*: www.dnzb.govt.nz.

How to Cook Paua
A.W.B. Powell, *Native Animals of New Zealand*: The Unity Press, Auckland, 1947, pp. 23–24.
Additional source: John Morton, 'Powell, Arthur William Baden 1901–1987', *Dictionary of New Zealand Biography*: www.dnzb.govt.nz.

The Rediscovery of the Takahe
Walter Mantell, from a letter to Gideon Mantell, 19 January 1850, Mantell Family Papers, MS-Papers-0083-341, Alexander Turnbull Library, National Library of New Zealand; Joan L. Telfer, 'Notornis Rediviva: A Mirror Nature Study', *The Mirror: New Zealand's National Home Journal*: vol. 28, no. 12 (June 1949), pp. 28, 29, 55–59.
Additional sources: J.R.H. Andrews, *The Southern Ark: Zoological Discovery in New Zealand 1769–1900*: Century Hutchinson, Auckland, 1986; William G. Lee and Ian G. Jamieson (eds), *The Takahe: Fifty Years of Conservation Management and Research*: University of Otago Press, Dunedin, 2001; M.P.K. Sorrenson, 'Mantell, Walter Baldock Durrant 1820–1895', *Dictionary of New Zealand Biography:* www.dnzb.govt.nz.

Takahe
Anna Jackson, *The Pastoral Kitchen*: Auckland University Press, Auckland, 2001, pp. 31–32.

Love-habits of the Yellow-eyed Penguin
L.E. Richdale, *Sexual Behavior in Penguins*: University of Kansas Press, Lawrence, 1951, pp. 15–27.
Additional source: Christopher Robertson, 'Richdale, Lancelot Eric 1900–1983', *Dictionary of New Zealand Biography*: www.dnzb.govt.nz.

The Third Man of the Double Helix
Maurice Wilkins, *The Third Man of the Double Helix: The Autobiography of Maurice Wilkins*: Oxford University Press, Oxford, 2003, pp. 118–212.

Making Waves: A Poem for Maurice Wilkins
Chris Orsman, *The Lakes of Mars*: Auckland University Press, Auckland, 2008, p. 22.

Establishment of a Carbon-14 Laboratory
T. A. Rafter, 'Problems in the Establishment of a Carbon-14 and Tritium Laboratory', *CONF-650652 Proceedings of the Sixth International Conference on Radiocarbon and Tritium Dating,* Washington State University, Pullman, Washington, June 7–11, 1965, pp. 752–61.
Additional source: Rebecca Priestley, 'Ernest Marsden's Nuclear New Zealand: From Nuclear Reactors to Nuclear Disarmament', *Journal and Proceedings of the Royal Society of New South Wales*: vol. 139, 2006, pp. 23–28.

The Allende Meteorite
Brian Mason and Simon Nathan, *From Mountains to Meteorites*: Geological Society of New Zealand, misc. pub. 109, Lower Hutt, 2001, pp. 56–58.

Premature Birth and Respiratory Distress
Graham Liggins, *A Brief Autobiography of Graham Liggins FRS* (unpublished); interview by Allan Coukell with Graham Liggins, *Eureka*, Radio New Zealand National, 3 February 2002.

Valley of the Dragons
Joan Wiffen, *Valley of the Dragons: The Story of New Zealand's Dinosaur Woman*: Random Century, Auckland, 1991, pp. 9–75.

Plastics that Conduct Electricity
Alan MacDiarmid, 'Autobiography', Tore Frängsmyr (ed.), *Les Prix Nobel, The Nobel Prizes 2000*: Stockholm, 2000: http://nobelprize.org/nobel_prizes/chemistry/laureates/2000/macdiarmid-autobio.html.

Bright Star
Christine Cole Catley, *Bright Star: Beatrice Hill Tinsley, Astronomer*: Cape Catley, Auckland, 2006, pp. 85–86, 153–54, 242, 373–74.

Sources

Lovemaps!
John Money, *Lovemaps: Clinical Concepts of Sexual/Erotic Health and Pathology, Paraphilia, and Gender Transposition in Childhood, Adolescence, and Maturity*: Irvington Publishers, Inc., New York, 1986, pp. xv–xvi.
Additional source: Michael King, 'The Duke of Dysfunction', *New Zealand Listener*, 4 April 1988, pp. 18–21.

A Survivor from the Dinosaur Age
Charles Daugherty and Alison Cree, 'Tuatara: A Survivor from the Dinosaur Age', *New Zealand Geographic*, April–June 1990, pp. 66–85.
Additional source: Charles Daugherty and Susan Keall, 'Tuatara', *Te Ara: The Encyclopedia of New Zealand*: www.TeAra.govt.nz/TheBush/ FishFrogsAndReptiles/Tuatara/en.

Tuatara
Nola Borrell, *Turbine 01*: www.victoria.ac.nz/turbine/borrell.html.

The Recent African Genesis of Humans
Allan C. Wilson and Rebecca L. Cann, 'The Recent African Genesis of Humans', *Scientific American*, April 1992, pp. 68–73.
Additional sources: Veronika Meduna, 'Allan Wilson, 1934–1991', *Atoms, Dinosaurs and DNA: 68 Great New Zealand Scientists*: Veronika Meduna and Rebecca Priestley, Random House, Auckland, 2008, pp. 90–91.

Ghosts of Gondwana
George Gibbs, *Ghosts of Gondwana: The History of Life in New Zealand*: Craig Potton Publishing, Nelson, 2006, pp. 74–75, 87–98.

Mismatch
Peter Gluckman and Mark Hanson, *Mismatch: Why Our World No Longer Fits Our Bodies*: Oxford University Press, Oxford, 2006, pp. 1–13.
Additional source: Veronika Meduna, 'Peter Gluckman, born 1949', *Atoms, Dinosaurs and DNA: 68 Great New Zealand Scientists*: Veronika Meduna and Rebecca Priestley, Random House, Auckland, 2008, pp. 110–11.

To Find a Planet…
Karen Pollard, 'To Find a Planet…', first published in this volume.
Additional sources: *Our Changing World*, Radio New Zealand National, 2 February 2006; 'Canterbury Astronomers Involved in Ground-breaking Planet Discovery': www.comsdev.canterbury.ac.nz/news/2006/060126a.shtml.

Copyright

Polynesian Navigation © Peter Adds, reproduced with permission of Awa Press; The Transit of Venus © Chris Orsman; 'Not a Pleasant Place' © Cambridge University Press, reproduced with permission; To Ernst Dieffenbach © Ian Wedde, reproduced with permission of Auckland University Press; Magnitude Eight Plus © Rodney Grapes, reproduced with permission of Victoria University Press; The Remarkable Pink and White Terraces translation reproduced with permission of the Estate of Charles Fleming; Huia © Anna Jackson, reproduced with permission of Auckland University Press; The Skeleton of the Great Moa in the Canterbury Museum, Christchurch © the Estate of Allen Curnow, reproduced with permission of Auckland University Press; Rutherford and the Discovery of Radioactivity © Cambridge University Press, reproduced with permission; A Most Incredible Event © Institute of Physics and IOP Publishing Limited, www.iop.org/EJ/journal/PR; Kakapo © Sonja Yelich and *Trout*; From Sea Floor to Coastal Plain, reproduced with permission of the Estate of Charles Cotton; Shell Collecting in the 1920s reproduced with permission of the Estate of Charles Fleming and the President of the Conchology Section of the Auckland Museum Institute; The Two Lucies reproduced with permission of the Auckland Botanical Society; Beginnings: Guthrie-Smith in New Zealand, 1885 © Peter Bland; A Man Who Moved New Zealand © Simon Nathan, reproduced with permission of Victoria University Press; Paradise Lost reproduced with permission of *Forest and Bird* magazine; How to Cook Paua reproduced with permission of Auckland War Memorial Museum; The Rediscovery of the Takahe reproduced with permission of Joan Watson; Takahe © Anna Jackson, reproduced with permission of Auckland University Press; Love-habits of the Yellow-eyed Penguin reproduced with permission of the University Press of Kansas; The Third Man of The Double Helix © Estate of Maurice Wilkins, reproduced with permission of Oxford University Press; Making Waves: A Poem for Maurice Wilkins © Chris Orsman, reproduced with permission of Auckland University Press; Establishment of a Carbon-14 Laboratory reproduced with permission of the Estate of Athol Rafter; The Allende Meteorite © Geological Society of New Zealand; Premature Birth and Respiratory Distress © Graham Liggins; Valley of the Dragons © Joan Wiffen, reproduced with permission of Random House New Zealand; Plastics that Conduct Electricity © Nobel Foundation 2000; Bright Star © Christine Cole Catley, reproduced with permission of Cape Catley Ltd; Lovemaps! reproduced by permission of The Kinsey Institute for Research in Sex, Gender, and Reproduction, Inc.; A Survivor from the Dinosaur Age reproduced with permission of Charles Daugherty, Alison Cree and *New Zealand Geographic*; Tuatara © Nola Borrell; The Recent African Genesis of Humans © Rebecca Cann and the Estate of Allan Wilson; Ghosts of Gondwana © George Gibbs, reproduced with permission of Craig Potton Publishing; Mismatch © Peter Gluckman and Mark Hanson, reproduced with permission of Oxford University Press; To Find a Planet... © Karen Pollard.

Every effort has been made to trace or contact copyright holders. The publisher will be pleased to make good in future editions or reprints any omissions or corrections brought to its attention.

Acknowledgements

Once I made the initial selection for this anthology, gathering the selected material into one manuscript became a technical, precision-driven process. Text from the original sources was scanned, put through an optical character recognition program, and the resultant text file was carefully read against the original text. My first thanks go to Ruth Brassington, who interrupted her own work to read each piece with me, word for word, along the way giving me her feedback. When she said, 'I love it!' I included it confidently; when she said, 'Boring as hell,' I ditched it and looked for something else.

I would also like to thank all the writers, and their publishers, whose work is included in this anthology, along with the families of deceased scientists and poets who gave permission for their relatives' words to be included. Particular thanks go to George Wilkins, who took the time to carefully read and respond to a very long extract from his father's autobiography, to Mont Liggins who gave me access to his unpublished autobiography from which I selected what I think is a fabulous piece, and to Karen Pollard, who wrote a piece especially for this anthology.

Further thanks go to Simon Nathan for suggesting I include an extract from Brian Mason's memoir, to Rowan Taylor for providing me with a selection of Allan Wilson's published articles, and to Sandy Bartle for responding to some queries about New Zealand's bird life.

Final thanks go to the wonderful team at Awa Press—Mary Varnham, Sarah Bennett and Hannah Bennett—and to my family, Jonathan, Pippi, Hazel and Huck.

Index

Page references in *italics* refer to illustrations.

Index

Index

Index

microFUN (Microlensing Follow-Up Network) 345
microlensing 343–47
microterigids 326–27
Milky Way galaxy 344–45, 347
minerals 15, 183, 257–58
Mismatch (Gluckman and Hanson) 328–42
missionaries 25–28
MOA (Microlensing Observations in Astrophysics) 345–46
moa 94, 210–11, 271, 297, 322–23
Moa's Ark theory 318–27
Moehau (Coromandel Peninsula) 153–55
Moeraki Headland 135
molecular clock 304–17
Molnar, Ralph (Dr) 273–75
Moncrieff, Pérrine (1893–1979) 179–92, 193
Money, John (1921–2006) 288–90
Monro, A.D. (Bobbie) (Mr) 279
Moore, Lucy (1906–87) 150–55, *152*
moraines 65–67
morepork 78, 90
mosasaur *267*, 268–69
mosquito 17
moss 57, 187
moth 69, 106–12, 326–27
Mount Maunganui 145
Mount Pender 129
Mount Tarawera 52, 95–105, *96*, *100*
Mundy, Daniel *58*
Murchison Range 199
Murrell, Norman 200
Museum of Natural History (Ottawa) 271
Museum of Natural History (Vienna) 83
mutations 307–13, 341
muttonbird 91, 93

N

Narbrough, John (Sir) 17
Nathan, Simon 177
National Institute of Genetics (Japan) 313
National Women's Hospital (Auckland) 260–61

Native Animals of New Zealand (Powell) 197–98, *198*
Native Bird Protection Society 179
native fauna *see* flora and fauna
native quail 182
native *see* Maori
natural selection 23, 68, 308, 315, 339–40
Nature 244, 342
Nauru Island (Pacific) 336–37
navigation 1–10, *7*
Nei, Masatoshi 315
New Caledonia 326
New Guinea 311–12, 314, 335
The New York Times 255
New Zealand Birds and How to Identify Them (Moncrieff) 179
New Zealand Geological Survey *see* Geological Survey
New Zealand government 123, 189
New Zealand Herald 146
A New Zealand Naturalist's Calendar and Notes by the Wayside (Thomson) 135–37
New Zealand Plants and Their Story (Cockayne) 138–39, 150
New Zealand Shell graduate scholarship 279
New Zealand wrens 322
Newman, Don 299–300
Ngapuhi tribe 71
Ngarauru tribe 169
Ngati Aurutu sub-tribe 164
Ngati Mutunga tribe 164
Ngati Ruanui tribe 169
Nicholls, Mary 227
Ninety Mile Beach 145
Nobel Prize 220, 246, 278
Norfolk Island 30
Norgay, Tenzing (Sherpa) 329
North Brothers Island (Cook Strait) 292
Norwegian rat 28, 180
Notornis Rediviva: A Mirror Nature Study (Telfer) 199–211
Novara 52
nuclear model of the atom 117–21
nuclear science 113–21, 250–54
nuclear weapons 220
nucleic acids structure 220–46

Index

Index